Silicon Chips and You

SILICON CHIPS AND YOU

The Magical Mineral in Your Telephone,
Calculator, Toys, Automobile, Hospital,
Air Conditioning, Factory, Furnace, Sewing
Machine, and Countless Other Future Inventions

■ C.D. Renmore ■

BEAUFORT BOOKS, INC.
New York   Toronto

Library of Congress Cataloging in Publication Data

Renmore, CD C.D,
    Silicon chips and you.

    Bibliography: p.
    Includes index.
    1. Integrated circuits.   2. Microelectronics.   I. Title.
TK7874.R46   1980   621.381'73   80-24153
ISBN 0-8253-0022-3

Published in the United States by Beaufort Books, Inc., New York.
Published simultaneously in Canada by Nelson, Foster and Scott Ltd.

Printed in the United States of America   First U.S. Edition

10   9   8   7   6   5   4   3   2   1

To Alison, Ian,
and all their generation

Silicon Chips and You

▪ Contents ▪

▪ 1 ▪
Introducing the Revolution

Everyone seems to have heard of silicon chips. The number of people who are familiar with silicon chips however, is probably equaled only by the number who have no idea (or, worse, the wrong idea) of what they are and what they can do. Is it just another pair of words that will disappear from the headlines as time passes?

This book is intended to convince you that silicon chips are very important. It is aimed at people who are wondering what all the fuss is about; wondering whether it really is worth the effort of finding out whether silicon chips should be cooked, rubbed into the furniture, or, preferably, just ignored.

Ignore them we cannot: We are already surrounded. Wristwatches, pocket calculators, television games, toys, and home computers are only the precursors, the advance guard. Do they represent an army of occupation or of liberation? These issues are, as I intend to show, of vital importance to every one of us.

You don't need to know anything about engineering or to have *any* scientific background to understand this book. On the contrary, curiosity and concern are the only two qualifications I ask of you.

FROM EVOLUTION TO REVOLUTION

Electronics can reasonably be said to have evolved from the first tube to the first transistor in about half a century. The first radios had a few tubes; the last computers had about eighteen thousand. I say the last computers because they very nearly were: Had the transistor not arrived in time (1947), computer engineering would have been hampered by the fact that finding and replacing defective tubes in more complex machines would have taken *more* than twenty-four hours each day.

Let us look at some round figures that give an indication of advances in transistor technology since about 1960. The figures below are the number of transistors that fit onto a chip of silicon about half a centimeter square:

1960: one	(1)
1970: one thousand	(1,000)
1980: one million	(1,000,000)
1990: one billion	(1,000,000,000)?

The question mark after the last figure refers less to the number itself than to the assumption that silicon will still be used to achieve such high packing densities. Some of the most interesting chips will not be made from silicon at all: The first superchips are already here (Chapter 4).

The growth rate underlying these numbers is a consistent doubling every year for (so far) two decades. It would be tempting, but naive, to extrapolate to the trillion level by the year 2000. Before that level could be reached, we will have to know how to split up the molecular building blocks of the materials we use.

These figures are impressive enough, but the decrease in cost and the increase in computing power that have accompanied these developments are further vital ingredients in the mixture that might be more precisely called the microelectronics revolution.

Consider cost first. Pocket calculator prices, for a given performance, have decreased a hundredfold in the past ten years, in spite of inflation. Electronic wristwatches with an accuracy that would have commanded the highest respect as well as the highest price from the traditional watchmakers of two decades ago are now among the cheapest as well as the most accurate of timepieces. The virtual extinction of the traditional watchmaking industry by this apparent outsider is a warning to us all.

Hand-held or desk computers, the equivalent cost of which would have significantly dented the budget of even a large corporation fifteen years ago, can now be purchased for a few hundred dollars. They have the capacity for processing information that, in those distant days, would have required a computer the size of a concert hall and the power to drive a steam engine. They are thousands of times more reliable, since there are no glowing filaments in the transistors as there were in tubes. As long as the temperature rise of the transistors is kept within set limits, they should long outlast their masters, working for centuries if necessary.

All this, of course, is still evolution, but it is so rapid that it amounts, almost, to a revolution in itself. There are other ingredients as well: Communications are themselves being revolutionized through the increasing use of glass fibers rather than copper wires. That is of vital importance, because only by the use of these fine strands of glass can information be sent between different locations at a rate compatible to that of the speed of modern computers (Chapter 9).

THE RULES HAVE CHANGED

Computer design has been an evolutionary process, starting when the basic building blocks, the transistors, were still rather expensive compared with the wires that connected them. Then came the method of joining whole arrays of transistors in a solid block of silicon, called an integrated circuit (Chapter 3). This has brought about a complete inversion of the economics: It is now the transistors that are virtually cost-free and the wires (in fact metal strip connectors on printed circuit boards) that are expensive.

This has important implications for the design of modern computers. The old theory of design set out to achieve a given computing function that involved the minimum number of transistors, which were expensive, and did not take into account the interconnections, which were cheap. New design theories must not only allow for this cost inversion, but also another factor that at first glance seems irrelevant: the lengths of the interconnections between the silicon chips. Why have the lengths suddenly become important?

The sad fact is that electrical signals cannot be made to travel faster than light. This sets a limit to the speed at which information can be exchanged between different parts of a computer. The speed with which modern transistors can operate (about one-thousandth of one-millionth of a second) means that the time taken for the signals to pass between them is slowing down the computer. Instead of thinking of the wires as joining the transistors together, we must think of them as keeping the transistors apart.

So, as well as considering the new economics, the designs must allow for the arrangement in space of the various parts. The fastest computers will therefore approach spheres in shape, or a set of interconnected spheres resembling the models of complex molecules seen in chemistry laboratories. To achieve the required surface area, it may be necessary to use the new geometry of fractals. These are spongelike objects whose internal surface area can be made indefinitely large for a given external size and shape. Their implications for computers are only just beginning to be appreciated.

The last and perhaps most spectacular convolution is concerned with size. No one would worry about a single speck of dust on one of the old radio tubes; but a dust particle can be larger than one of the transistors on a silicon chip. The role of the dust particle has been transformed from insignificance to dominance. Such contamination can cause the chip to fail.

The computer is gradually disappearing, not because it is no longer there, but because it is becoming too small to see.

The new rules of the game have yet to be fully formulated, but here are some simple guidelines:

1. We now have a virtually infinite capacity for the storage, processing, display, and communication of information.
2. The cost of the electronics is approaching zero.
3. The speed of the electronics is increasing toward limits set by the speed of light.
4. The size of the electronics is decreasing toward limits set by the granularity of matter itself.

There is a fifth axiom that is really contained in the first of these rules, but which is of such tremendous significance that it requires separate discussion (See Chapters 2 and 8). It concerns the emergence of the hyperintelligent machine (HIM).

I cannot forecast the full consequences of even these simple guidelines, but they are bound to be impressive. We can perhaps see the nature of the problem by cheating and applying hindsight to another somewhat similar challenge.

Consider surveying, aircraft design, and interplanetary navigation as just a few examples of precision engineering that use the results of geometry; the geometry of Euclid. The whole of that geometry, including the famous Theorem of Pythagoras, can be worked out from just a few rules and definitions called axioms. The axioms refer to points and lines, nothing more. Now imagine yourself, two thousand years ago, formulating the axioms. It would be a mighty achievement to anticipate a few of the major theorems of geometry, let alone foresee what extensions of our abilities they might ultimately help to bring about. That is the nature of the challenge, except that we are concerned with the next twenty years rather than the next two thousand, such is the pace of the revolution.

While on the subject of what it all might mean, I should mention that the arrow of time does not always fly straight. Let's take a simple question first: We all do arithmetic at school, but what do professional mathematicians do? If you ask someone who did only arithmetic before leaving school, he will recollect that he had to multiply and divide numbers, then larger numbers . . . and he will probably say that mathematicians are people who multiply and divide absolutely enormous numbers, occupying sheets and sheets of paper. They may use calculators or even computers, but the central idea is one of sums. It would not occur to most people that mathematicians might be more preoccupied with things that replace concrete numbers by abstract symbols, that is, with algebra and beyond. Can we blame someone for thinking along these lines? It is perfectly natural and logical to build in this way on known foundations. Nevertheless, it can lead us astray. Now for a second example, much closer to home.

Suppose that we traveled thirty years back in time to consult the famous Swiss watchmakers about their latest chronometers. We ask them the question, How long before your chronometers are one hundred times as accurate, yet smaller and lighter as well? Inevitably, their minds would turn to the problems of refining existing techniques: better balance wheels, bearings, temperature compensation, even perhaps new materials. But never silicon. It wouldn't have made sense. And if we had asked that the new watches should have no moving parts, I doubt that we would have been taken seriously. Well, we know what happened: The mechanical watch has been superseded by a totally different conception. The arrow of time took an unexpected direction, as it may do again in the present revolution. The hazards of prophecy are delightfully expounded in a book by Arthur C. Clarke called *Profiles of the Future*, still undated after almost twenty years, and in his *Report on Planet Three*.

It is my object in this introductory chapter to convey the essence, the flavor, of the revolution rather than to dwell upon details that will be found later in the book. So here are some glimpses of things to come. You will, I am sure, forgive a certain vagueness in parts. Memories of the future are necessarily indistinct.

SOME MEMORIES OF THE FUTURE
The car wouldn't start this morning: It didn't recognize my voice. To be fair, I must admit that I had a cold and couldn't get through the identification sentence without a snuffle or a cough, and that must have made it suspicious. I got my own way eventually, though, by trying a trick that would never work on the latest models. I played a tape to it with the sentence word-perfect. It started first time.

You can get one-word starters, but the insurance premiums go up so much that most people prefer the tongue twisters like the one mine has. The point, of course, is that you could never get through that if you were drunk.

My car isn't the brightest, but it keeps me out of trouble and that's all I ask. Any necessary maintenance or impending faults are clearly spelled out on the dashboard display (some models tell you what's needed; some even contact the garage). If I ignore the warnings and have an accident, I'm permanently disqualified from driving. It's all recorded in the black box, you see, like the ones they use on aircraft.

Television isn't what it used to be. There are no peak viewing hours or difficult decisions about what to watch, since all the programs (well, almost) are just dialed. You watch when you want to, in other words.

The children were responsible for that. They have their homework and their hobbies, and the teaching channels. . . . Well, I must admit that I'm impressed. They don't just talk back, those machines, they adapt to the ability level of the person they're teaching. Almost uncanny, but very effective. And always so encouraging, such tireless enthusiasm and infinite patience. The kids love them. I keep reminding them (or am I just trying to convince myself?) that they are only machines. All I get back is, "Don't worry, Daddy, we love you too."

I think what really converted me was that their teddy bears were not just better at arithmetic than I am, but they could explain it so much more clearly, and with such patience and good humor, too. And in so many different voices, until they get to the one that the children like most. The toys won't talk to me, by the way, which is bad for my ego. My conversation with the bears is limited to a polite request for a set of new batteries before they start getting their sums wrong, or forget the children's names.

Yes, it was the children, the young people, who led us into the revolution. It's not so surprising, really. Most adults, including me, are suspicious and resentful of computers and robots for all sorts of reasons, but most children have always loved them. Where we saw our jobs threatened, they saw only release from the drudgery that my generation called "jobs." It makes sense; the children were right. After all, *they* are going to inherit the Earth, aren't they?

• 2 •
Silicon Chips:
What They Are and
What They Can Do

A silicon chip is a piece of almost pure silicon, usually less than one centimeter square and about half a millimeter thick. It contains hundreds of thousands of microminiature electronic circuit components, mainly transistors, packed and interconnected in layers beneath the surface. On the surface of the chip there is a grid of thin metallic strips; these are electrical connections via wires to the outside world. When sold commercially the chips are generally packaged in plastic.

There are several types of silicon chip. A *microprocessor* chip is the part of the computer that controls and performs all the arithmetical and logical functions required. Another type of chip, a *memory* chip, stores information in such a way that it is readily accessible to the microprocessor chip when it is needed.

There are two kinds of information storage chips, *data* memory and *program* memory. A data memory chip stores raw information upon which operations are to be carried out. An example would be a list of numbers that must be put into a formula to work out the fuel consumption of a car. These figures would differ with each design of car, and the data memory chip can store numbers corresponding to many different types of vehicles.

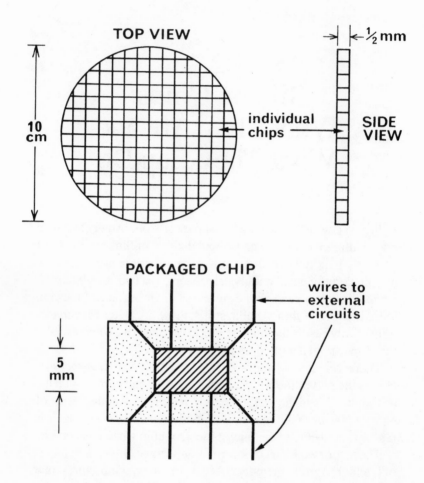

Each chip contains a complete microelectronic integrated circuit. Several hundred can be made on the same silicon wafer.

The chip is mounted in a plastic package with external connecting pins. Many hundreds of packaged chips are mounted on a circuit board and interconnected to form part of a computer.

A program memory chip stores a set of instructions specifying the stages in the calculation from the formula and the sequence in which they must be carried out. Referring to the example of fuel consumption in a car, the numbers stored in the data memory will be replaced by new ones as new designs of cars are considered.

A third type of chip is a *communications* chip that enables the computer to interact with the user, for example via a keyboard and a television screen. It also enables the computer to gain access to other information stores (such as magnetic tape) and to other computers.

It is quite easy to put a computer on a chip. A complete computer on a chip incorporates the functions of a microprocessor, data and program memories, and a communications chip. It is known as a microcomputer.

The functions of a computer are to store, process, and display information. The information might be in the form of stock lists for a chain of supermarkets. The computer would store all the current stock details. The processing might consist of comparing existing stocks with likely future demand and noting potential shortages in certain branches. The display would then consist of a concise presentation of the conclusions so that the ordering department of the chain of supermarkets could plan its bulk purchases for the following months.

The computer on (and in) a chip consists mainly of *transistors*. A transistor is basically just a switch that can operate in a very short time. Transistor switches can turn themselves on and off in one-thousandth of a microsecond (a microsecond is one-millionth of a second).

Now, let's concentrate on what a switch is. The light switch that you operate by hand is a good starting point, since we are all familiar with it. It consists of two wires that either are or are not touching. If they are touching, then electric current can flow, and the switch is on. If not, it can't and it's off. There are only two possibilities, only two possible conditions for the switch; that simple fact is vital.

Assuming that there is nothing wrong with the light bulb or the power supply, you can tell whether the switch is on or off by checking whether the light is on or off. In everyday life our main interest is in the light, not the switch; in computers, the vital information is dependent on whether the switch is on or off. In computer terminology the very least amount of information that you can have that is of any use is called a *bit* (an abbreviation of *bi*nary digi*t*). If you begin by not knowing whether the switch is on or off, and you end by knowing that it is definitely one or the other, then you have gained one bit of information in the process.

Everything follows from this, strange though it may seem. All I can do here is to bridge the gap, very briefly, between the simple switch and the glimmerings of computer intelligence.

We all know about *decimal* arithmetic because we use it in our currency as well as in working out ordinary sums—on a calculator, of course. Though the answer comes up in the familiar form, the calculator (or computer) does not carry out the calculations using decimal arithmetic. It uses only the two numbers 1 and 0, which is called *binary* arithmetic, and then it converts the resulting string of 1's and 0's into a string of numbers, such as 3.14159. In other words, it uses a wider range of symbols when it communicates its answer to you. But *the information content is identical* in the two cases, whether the result is expressed using only two basic symbols or using ten. The main difference is that binary arithmetic is more economical.

Binary arithmetic is in many ways much simpler to deal with than decimal arithmetic, once you are used to having only two available numbers, 1 and 0, to represent all others. In particular, the result of adding two numbers *must* involve, again, only these two. Further, all the other operations in arithmetic (subtraction, multiplication, and division) can ultimately be reduced to addition. We are now ready to take the first real step toward computer intelligence.

Switches are either on or off; binary digits are either 1 or 0. So we can decide, arbitrarily, that a switch that is on represents a 1 and a switch that is off represents a 0. Then, by making switches that can turn one another on and off, we have the possibility of constructing a machine in which the answer to a calculation appears as a row of switches, their positions (either on or off) giving the final numerical answer in binary form. Thus, a row of switches (which are on or off) can represent a row of numbers (which are the numbers 1 or 0). A computer can manage without any special circuits for subtraction, multiplication, and division provided that it can add. What this system lacks in elegance it makes up in simplicity and speed. Millions of individual operations can be carried out in one second, and this will be accelerated a hundredfold when the superchips are in widespread use (Chapter 4).

Not only the numbers themselves, but also the instructions for how to deal with them, can be put into this same binary form. This has profound implications and is the second step toward computer intelligence. The instructions that are stored in a computer are called its *program*; the numbers the program manipulates are called *data*. One computer can alter the program instructions in another computer *or even in itself* (computers can learn from experience), as well as being able to manipulate data (do sums).

The idea that all computers can do is work out sums is perhaps at the root of the popular conviction that they are just high-speed morons and certainly not intelligent. Well, it just isn't so, and that brings us to step three. Over a century ago the mathematician George Boole published a paper with the title ''The Laws of

Thought.'' He set out, brilliantly and definitively, the ground rules for a system of symbolic logic. He developed a form of shorthand that enables complex logical problems to be handled as though they were exercises in algebra.

How does this affect the switch? Switches can be thought of as representing 1 or 0, on or off, high or low, yes or no, *true or false*. The algebra of Boole contains just these two symbols. It is binary in character, and perfectly adapted for computers to handle.

The part of a computer that stores bits of information is called the *memory*. A string of switches with their pattern of 1's or 0's is a convenient way of storing numbers and anything that can be represented by numbers. Consider that last point, that a computer can manipulate numbers or *anything that can be represented by numbers*. This hints at the immense power of the computer— the power of simulation, the power to represent one world in terms of another. Numbers can represent lengths, weights, shades of color, speeds, or television pictures. The list is endless.

The other main part of the computer is the part that organizes the information so that it is available when needed, carries out the instructions, and generally keeps things ticking over in the correct sequence. *Microprocessors* carry out this function; the microprocessor chip in a small computer is often vastly outnumbered by memory chips, but this balance may easily change as more compact computer memories are made (Chapter 4).

I hope to convert you to the view that computers are intelligent. It is my privilege to choose the battleground, and I choose to measure the human intellect against that of the computer in the subject that is generally agreed to be the most demanding intellectual game ever devised. It is a game of war, waged at the highest level; it is beyond the domain of the generals and enters that of the grand masters: I refer, of course, to international chess.

THE MIND OF THE MACHINE

Intelligence is very hard to define, but easy to recognize. Here are some of the characteristics that tend to go with our idea of intelligence: the ability to learn from experience, the ability to make decisions based upon information, and the ability to *solve problems*. If we fasten our attention on this last point, we might agree that, within a given field of problem-solving, the more difficult the problems you can solve, the more intelligent you are in that particular field.

Computer chess was a joke in 1957 when the first chess-playing program was written. All but the weakest club players could beat it. A good historical analogy might be the delight of those who held gas shares on the stock market when they learned that some of the newfangled electric lamps invented by Edison failed after only a few hours. It simply, and incredibly, did not occur to them that the next generation of electric lamps just might be better.

In 1967 a computer chess program was developed that could put up a more creditable performance. However, the real shock came when the program discovered something that had been overlooked by several chess masters, something called a sacrificial continuation. This is a move in which you deliberately lose a piece in order to be able to launch a strong attack. The publicity that accompanied this discovery prompted a lot of computer programmers to think about chess, and a lot of chess players to revise their opinions about computers.

In 1978 an international chess master and former computer programmer, David Levy, won a competition against a computer, but he did not win overwhelmingly: one game drawn and one game lost against the computer in a total of five games.

The earliest chess-playing computer could beat perhaps 10 percent of the world's chess players. Present computers can beat 99.9 percent of those players; even the international grand masters are not safe. Yet the chess-playing computers, for all their intellectual power, occasionally show some endearingly human characteristics: One such computer, as its very first move in its very first game, chose to resign.

The computer can be "better" than the person who programs it. A first-class athletic coach may not himself be an Olympic gold medalist; similarly, programmers can indicate what should be done and how it might then be done better. They do not necessarily have to be able to do it themselves. They inculcate the foundations of the subject and the ability to learn and adapt in the light of experience.

This is well illustrated by the present chess computers, which can almost invariably beat their own programmers at the game. This point is hard to grasp, yet it is crucial for an understanding of what intelligent machines are going to mean for our society. The idea that "a computer can only do what it has been programmed to do" contains the seeds of its own destruction. My reply to this objection is: Quite so, but the program can include instructions for learning and adapting, for absorbing information from the outside world, and for basing decisions upon that information. It can contain a general framework of instructions within which flexible and adaptive behavior is possible. An example of this kind of programming is evident in humans. The genetic program in every unborn child would have turned us into a race of automata long ago, had we not been provided with an adaptive, flexible set of genetic instructions.

That is the fourth step toward computer intelligence: to learn and adapt on the basis of experience. Chess-playing computers can undoubtedly do this. They improve with every game, and the better the opponent the more they learn. Soon they will only be playing chess with one another, or perhaps they will invent something more challenging that we will not even be able to understand. Chess championships in which both contestants are computers are now common. Soon we shall see chess doubles championships, in which each team will consist of one human and one computer. The strategic genius of the grand master will combine with the infinite memory and analytical speed of the computer to produce a uniformly brilliant level of play such as the world has never before witnessed.

Present computers examine a few million positions before making a move; before many years pass it will be billions and then trillions of positions. This brings me to the test of computer intelligence; only one test, admittedly, but in the context of this discussion one is enough to make the point.

The test was first suggested more than a quarter of a century ago by an English mathematician, Alan Turing. In its original form it was a test of general intelligence. In the context of chess, it would be somewhat as follows. Suppose that a chess grand master agreed to play against an opponent he could not see. If, from the quality of the opposition, he could not tell whether the opponent was a human or a computer, and the opponent was indeed a computer, then the machine must be considered intelligent.

Whether this precise test has been carried out yet I do not know, and it does not matter. The fact that international grand masters have occasionally lost to computers is enough to establish the existence of computer intelligence.

THE COMING OF THE HYPERCOMPUTER

In this last section I will outline some of the implications of computer intelligence, although these will be discussed more fully later on. The strengths of present computers lie in their capacity for high-speed calculations, logical analysis, and in such areas as information retrieval and pattern recognition. Here are a few ways in which they can help us.

Complete libraries and, indeed, all documented human

knowledge could quite easily be stored in a roomful of optical-memory devices (Chapter 4). The information implosion will continue, and in due course the documents themselves will be formed in the computer in the first place. The resources available to people studying at home will be essentially unlimited, with considerable consequences for education.

Speech synthesis and recognition, already well advanced, will be still further facilitated by the advent of microminiature computers. Reading to the blind, conversational computers, and language translation will benefit tremendously.

Can computers help us with our social problems? I have my doubts, but they can certainly help us with our material problems, if we will only let them. The operator of a mechanical digger does not feel inferior because the machine he controls has a thousand times his own strength. On the contrary, I imagine that it is an exhilarating feeling. Engineers are already consulting computers about design problems that they cannot solve by themselves (Chapter 3). Indeed, they have been doing so for many years. The pace of the revolution is accelerating now because the computers are intelligent enough *to design their own successors.*

Are you afraid of robots? A robot is a computer that can interact physically as well as intellectually with the outside world. They can do all sorts of jobs that are dirty, dangerous, and altogether unsuitable for their human masters.

Perhaps the thought of being in the presence of superior intellects is what worries most people about computers. In that case it seems to be only a matter of education, and our children are already enlightening us—just look in the toy shops.

• 3 •
Silicon Technology:
Some Glimpses

In Chapter 2 we saw that the power of the silicon chip has its origins in the fact that it is a microminiature computer. Now it is time to take a brief look at the background technology. How is it possible to put a million transistors on a slice of silicon only half a centimeter square? How can the cost of these electronic switches continue to halve every few years when the cost of everything else seems to double in the same period?

To answer these questions, I am going to use some unfamiliar words. Micrometer and microsecond are two whose meaning must be clear before I continue.

In silicon chip engineering, a distance of one meter is so enormous that it is almost equivalent to infinity, so the unit used instead is the *micrometer*, which is one-millionth of a meter or one-thousandth of a millimeter.

Similarly, one second is so near to eternity that we don't use seconds either. We use *microseconds*, that is, millionths of a second. As a matter of fact, even these units are starting to become outdated.

In this chapter I will describe how these tiny electrical circuits are made and how some of them work. In Chapter 4 we will move on to some chips that do not use silicon. At present these serve to supplement the existing silicon technology; eventually they will replace it, only to be replaced in their turn.

It will help to build up a picture of the silicon chip revolution if I delve briefly into the history of electronics. The theory is simple: If you know where it comes from and where it is now, you stand a better chance of guessing where it will be tomorrow.

TOWARD SPATIAL INTEGRATION

The transistor was invented in 1947 at Bell Telephone Laboratories in the United States. Even in its crude form, its small size in comparison with the old radio tubes that it was destined to replace was obvious. The transistor was dwarfed on the laboratory bench by the equipment they were using to test it. The possibility of a really compact, complete circuit (a radio receiver, for example) led the engineers to start dreaming about the possibility of something they called a *solid circuit*; and when engineers start to dream, things start to happen.

They envisaged a circuit in which all the electrical components were somehow built together in the same block of material instead of being made individually and then soldered together using wires. By 1960 the *integrated circuits*, as we now call them, were in existence. Solid state technology was born; the expression ''solid state'' signifies that the electrons are traveling through solid material rather than in a vacuum as with the old tubes.

The solid material from which these integrated circuits on chips are made is now usually silicon, though the first transistors were made using germanium.

Silicon, which is a major constituent of common sand and therefore an abundant element, belongs to an important class of materials whose electrical properties are intermediate between those of good conductors (like the copper wire used in domestic lighting circuits) and good insulators (like the plastic sheath around the wires). They are known, with uncharacteristic good sense, as *semiconductors*. Other semiconductors include gallium arsenide, gallium phosphide, indium antimonide, and indium phosphide. Some surprising substances turn out to have semiconducting properties: diamond, for example, which is a form of carbon. Gallium phosphide is used to make the microscopic lamps found in many calculator and wristwatch displays. Gallium arsenide can be used to make a laser on a chip—something that silicon cannot do (Chapter 4). Gallium arsenide technology will overtake silicon very soon, because it can be used to make faster computers.

Semiconductors are remarkable materials, provided you can get them in a pure state. The problem of unwanted impurities obscured many of their most spectacular properties before the invention of the transistor.

Perfectly pure silicon is a poor insulator, but its electrical properties are completely transformed when minute amounts of either boron or phosphorus are added to the structure of a perfect crystal. In fact, if a two-layer structure (called a *diode*) is made in which one layer has boron added and the other has phosphorus, an electrical component of great value is formed: It will allow electric current to flow through it in one direction but not in the other.

A three-layer sandwich structure in which the outer two layers are the same is one form of transistor. The transistor is of great interest to us because it can amplify and it can behave as a switch. We considered switches in Chapter 2; now we must consider how transistors *amplify* and then how they can be made to act as switches.

The basic idea of amplification is very simple. All that is needed is a system or a device in which a small change in one thing produces a much larger change in something else as a direct result. Much earlier generations than ours knew about amplifiers: A small tug on a ring through the nose of an ox produced a very much larger tug on the plow from the harness, and that amplifier evidently worked very well.

Now we can consider how the two main types of transistor amplify, and how that property can be used to make them into switches. Switches, remember, are the building blocks of computers.

FROM AMPLIFIER TO SWITCH:
EXTREME MEASURES

The important thing about a switch is that it is either on or it is off; there must be nothing in between.

In the previous section I referred to a transistor made in the form of a three-layer structure. We can see how this can amplify by considering water flowing over a dam from a lake. Suppose that the flow of water is regulated by means of sluice gates whose height is controlled very accurately by a handle. The height of this barrier will determine the flow of water from the lake; in particular, a small decrease in the barrier height will cause a large increase in the flow. This is the essence of amplification, and when we use it only between extremes, it becomes a switch; either there is maximum flow or there is none. The flow of water in this simple example corresponds to the flow of electricity in the transistor. It is an electrical switch that is itself electrically operated.

In the other kind of transistor, the amplification can be understood if we think of water flowing through a hose. Squeezing the hose constricts the channel available for the water and regulates the flow. As with the first analogy, a small change in the pressure put on the hose will make a large difference to the flow of water; and a hard squeeze will stop it altogether. By taking extreme measures, we can turn an amplifier into a switch.

It is quite possible to construct a computer from any sort of arrangement that will flip from on to off. For example, there is a science known as fluidics, part of which is the performance of logical (that is, computing) operations using jets of liquid that will switch from one arm of a forked pipe to the other. The problem is one of speed. In pursuit of the highest possible switching speeds, we cannot consider machines with moving parts that are any more clumsy than electrons (the carriers of electric current) or photons (the carriers of light).

The transistors I have described are at present able to switch from on to off in times ranging from about one microsecond down to as little as one-thousandth of a microsecond.

CHIP TECHNOLOGY:
CLEANLINESS EVEN BEFORE GODLINESS
I mentioned earlier that pure silicon is necessary. Well, it doesn't have to be absolutely pure: 99.9999999 percent will do. That is not an exaggeration. If more than one part in a thousand million is impure, the final yield of working chips is likely to be too low to be economic. Any process that can produce that kind of purity has to be something special; yet the inventor of the method, William Pfann at Bell Telephone Laboratories in 1938, did not publish his discovery until ten years later. The process was so simple, he assumed everyone knew about it already.

He made his discovery by noticing that when silicon ingots were withdrawn slowly from the furnace (so that they solidified gradually from one end) the impurities tended to collect at the end of the bar that solidified last. It turned out that the impurities, faced with an advancing wall of solidifying silicon, tended in most cases (not all) to maintain a higher concentration

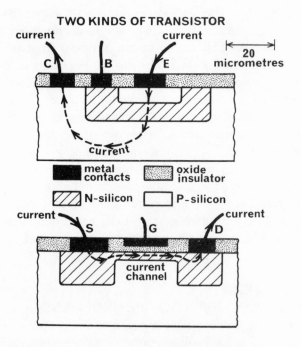

Top: *This kind of transistor, called a bipolar transistor, controls the current flow between terminals E and C in much the same way as water flow over a dam is controlled by the height of the sluice gate. It is a three-layer (N, P, N) structure where N-silicon means impregnated with phosphorus, and P-silicon means impregnated with boron.*

The height of the electrical barrier to current flow is determined by the voltage between terminals B and E. There can be one million of these transistors in a single chip. (Terminal E is called the emitter, B the base, and C the collector.)

Bottom: *This is called a field-effect transistor. The current flow between terminals S and D is controlled by the voltage applied to terminal G. This controls the thickness of the channel and regulates current flow in much the same way as water flow in a hose can be controlled by squeezing the pipe. (S is called the source terminal, G the gate, and D the drain terminal.)*

in the liquid than in the solid/liquid boundary. The result was that a bar with an initially uniform distribution of impurities came out of the furnace with most of them swept to one end; that end was promptly sawn off and the process repeated. Soon it was found that the whole sequence could be made more automatic by using traveling molten zones.

Instead of melting the complete silicon bar, only a short length close to one end is melted, using a heating coil. By slowly moving the coil along the length of the bar, a molten zone traversed its length, carrying a useful proportion of the impurities with it. As soon as the first coil was far enough away from the starting point, a second coil produced a second molten zone to follow it. In this way, several sweeps of the bar could be carried out simultaneously and the whole process repeated until there were no further reductions in the impurity content. This elegant method, known as zone refining, is a cornerstone of silicon chip technology. Without it, many of the devices I have described might still not work on silicon no matter how well they work on paper.

Zone refining is by no means limited to the purification of silicon. It has been successfully applied to the purification of many hundred organic compounds and about a third of all known elements.

The level of purity we are considering here is difficult to retain once it has been achieved. There is the problem of contamination from the atmosphere, which can be controlled during the zone-refining process but not so easily afterward. Special clothing is worn, the air is filtered, and the dust count is at least ten times lower than in the best hospital operating theater, but still the contamination creeps back. The worst danger of all is from the container in which the silicon has to be supported while the molten zones pass along it. Quartz containers are among the least contaminating, but the problems inherent in keeping the

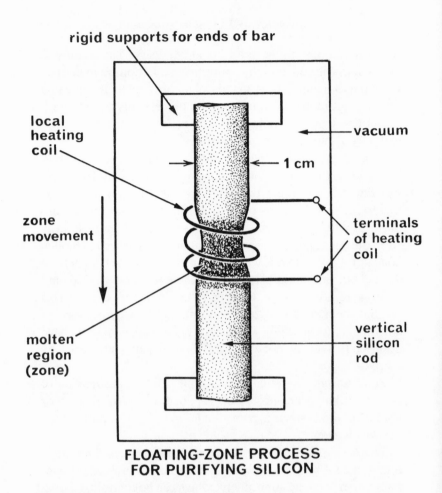

**FLOATING-ZONE PROCESS
FOR PURIFYING SILICON**

Zone Refining: *This method of purification is a cornerstone of silicon chip technology: There is no other known means of purifying silicon sufficiently (less than one part in ten thousand million of impurities). Molten silicon is at 1400° C. and at this temperature it would react with all known container materials, becoming contaminated. A method had to be devised of allowing a molten zone to travel down the rod, sweeping the impurities with it, yet not needing a container.*

silicon pure led to an extreme measure. Molten silicon is at a temperature well in excess of 1,000° Centigrade, yet it was decided that there should be *no container*.

This form of the process, called the *floating-zone* method, has a vertically mounted silicon rod down which a molten zone travels, controlled by a heating coil. The zone is not easy to stabilize, it is only kept from spilling by its surface tension, but it can be done. The resulting silicon is about a hundred times more pure than the limit set at the beginning of this section.

The silicon is now sufficiently pure, but it is not a single crystal, and these are essential. To produce a single crystal, the zone-refined silicon is melted in a crucible and a small seed crystal, as near perfection as possible, is dipped into the melt. The seed crystal is then slowly withdrawn. Using the perfect lattice of the seed crystal as a template, the silicon in the crucible can be pulled out as a single crystal which can be ten centimeters in diameter and up to two meters long. It is at this crystal-pulling stage that the minute quantities of either boron or phosphorus are added to the melt.

The crystal is next cut into slices, called wafers, using a diamond-edged circular saw; the wafers are generally about half a millimeter thick. One side is then highly polished. There must be no scratches, no defects. The subsequent processing the wafer must undergo depends upon a scratch-free surface that is uniformly flat to within one wavelength of visible light. Hundreds of wafers come from a single crystal, and hundreds of silicon chips will eventually be produced from each wafer. Add to this the fact that a large production plant can process tens of thousands of wafers every week and we can see how it is possible to produce silicon-chip devices so cheaply. Furthermore, it is now clear why compact circuit design is of such crucial importance: More chips per wafer mean cheaper devices.

We are now poised and ready to make a microelectronic circuit on a silicon chip, except for one thing—we will have to design the circuit.

THE SIMULATION GAME:
SET A CHIP TO MAKE A CHIP

There was a time when, if you wanted to try out a new idea for a circuit, you took the components and soldered them together on something known in the trade as a bread board. If the circuit didn't work, it was not too much trouble to change a few components and retest it. There are one or two difficulties in applying this approach to the design of a silicon-chip microcircuit.

First, a microscope would be needed to see the components; and there would be no question of being able to handle them individually in any case. Second, the design may involve a million circuit elements, a change in any one of which will affect the response of every one of the others to some extent—and how can we decide to what extent?

This is rather alarming: It is an open admission that no human being can, unaided, either design or produce a microprocessor on a silicon chip. Yet, and this is my point, it should not be at all disturbing to make this admission. Human technology is a progression of crude tools being used to fashion better ones; now we can summon intelligent tools to help us make more intelligent ones, and sharpen our own intellects into the bargain.

The basic conception of the design is supplied to a computer together with approximate positions and sizes of all the circuit elements involved. Then the behavior of the circuit is calculated in detail by a circuit-analysis program, which also tells the designer which parts are going to have the greatest effect if they are changed. The numerical value or physical position of any component can be altered by typing in a correction; the effect of this change on the overall behavior is then given immediately on a display screen. The circuit designer is interacting with a computer to design that computer's successor.

Any part of the layout can be displayed, magnified several hundred times, on the screen. The circuit is not simply going to be a flat pattern on the silicon surface, it will be made up from several layers, all interconnected. The problem is to achieve all the specified circuit functions in the minimum possible area on the silicon surface.

As the final design is approached, the computer produces drawings that are checked manually for accuracy. Since the individual lines on the final circuit may only be a few micrometers apart, these drawings are produced at about one thousand times full size so that they can be inspected easily. Nevertheless, it is a formidable and time-consuming task. The stakes are high: One error on one part of the circuit may be multiplied a thousandfold in production before it is corrected; and one error either in design or in execution is likely to be enough to make the whole circuit fail on the first electrical test.

A particularly difficult design, for example, a new microprocessor circuit, may take years to complete. More routine and repetitive designs can take as little as a few weeks.

When the circuit is designed, the task of impressing this minute and intricate pattern upon the silicon wafer is faced. The first stage is photographic reduction.

REMEMBER THE MICRODOT?

The microdot, favorite tool of spies (at least in fiction), is a reduced photograph of a complete document, small enough to escape notice on a printed page. To read the document, you need a microscope. Imagine an array of microdots, each containing a circuit diagram, and you have the basic idea of microcircuit fabrication. The process is known as *photolithography*. To impress the pattern of the circuit on the silicon we need a microscopic stencil, called a *photomask*, for each layer of the circuit. Our next step is to consider one of these stencils.

The computer has the exact layout of the circuit stored in its memory, and it prints an image of the circuit, magnified by ten, onto a photographic plate. The print is produced by scanning a spot of light, controlled by the computer, across the plate. From this original, an image that is the correct final size of the circuit (typically this might be half a centimeter square) is reproduced hundreds of times in rows and columns to fill an area the size of the silicon wafer. This final print is the photomask or stencil for the particular layer of the microcircuit.

The final circuit will consist of layers precisely positioned above and below one another. This means that the alignment of the successive photomasks on the surface must be very precise, accurate to within about one micrometer over the whole area of the slice.

When the circuit is complete, some of the layers will be above the level of the original silicon surface (the metal connecting strips that serve as wires, for example) and some will be below it (the sandwich structures that form transistors and diodes).

To keep the description as simple as possible, I will not attempt to describe the stages in the production of a complex integrated circuit. The essential ideas involved can be conveyed by considering a much more simple structure. Suppose that we want our circuit to consist simply of a regular pattern of diodes (the two-layer structures mentioned in the section on spatial integration) buried beneath the silicon surface, and accessible by metal contacts above them along the surface of the wafer. The stages involved in this are representative of the stages of the fabrication of much more complex circuits.

SETTING THE PATTERN

Suppose that the wafer had been originally impregnated with boron; the pattern of diodes must then involve the introduction of phosphorus to a controlled depth on selected areas of the wafer only, namely those areas where we want the diodes to be. The diodes will then be islands of phosphorus-impregnated silicon in a pattern defined by the photomask. This is achieved by first covering the whole silicon surface with an insulating layer and then opening windows in the insulator through which the phosphorus can be introduced.

One of the main reasons for the popularity of silicon as a material for integrated circuits is that it forms an oxide (silicon dioxide, chemically very similar to sand) on its surface. Silicon dioxide is one of the best insulating materials known. Unlike the oxide that forms on the surface of iron (rust), silicon dioxide does not flake off, but adheres very well. The oxide layer is formed on the surface of the wafer by putting the wafer in a furnace at approximatley 1,000° Centigrade in an atmosphere of pure oxygen. The wafer itself is inside a long quartz tube and the heat comes from external heating elements to minimize contamination of the surface.

The oxide grows slowly, allowing great precision in the control of its thickness. For example, under the conditions described, the layer will form at the rate of one-tenth of a micrometer per hour. Many hundreds of wafers can be processed at the same time, and the temperature of the furnace is regulated to within a fraction of a degree by continuous computer control.

The oxide is formed when the surface of the silicon reacts with the oxygen surrounding it, so that the first micrometer or so of the silicon is converted into insulating material together with any surface contamination it may have acquired. Because the oxide is such a good insulator, these impurities are surrounded,

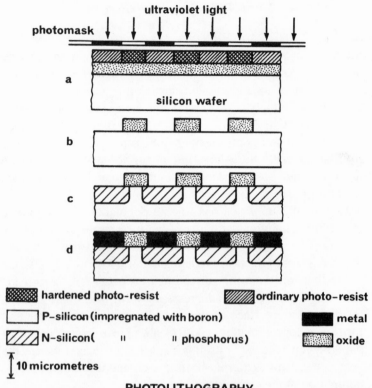

ultraviolet light

photomask

a

silicon wafer

b

c

d

hardened photo-resist		ordinary photo-resist
P-silicon (impregnated with boron)		metal
N-silicon(" " phosphorus)		oxide

10 micrometres

PHOTOLITHOGRAPHY

a. *Ultraviolet light is shone onto the light-sensitive surface through the photomask. Where the light falls, the photo-resist hardens. The pattern of this is the pattern of one layer of the final microcircuit.*

b. *The ordinary photo-resist and the oxide below it are dissolved, then the hardened photo-resist also, leaving islands of oxide in an otherwise exposed silicon surface.*

c. *The wafer is then put in a furnace at about 1000° C. and exposed to phosphorus vapor. The vapor diffuses into the silicon surface and partly under the oxide. This forms an array of two-layer structures called PN diodes.*

d. *Finally, metal contacts are evaporated in the form of a thin film so that wires can be bonded onto the chip when it is packaged.*

trapped, and electrically isolated. This goes a long way toward making up for the various inevitable sources of surface degradation since the silicon was first zone-refined. A further bonus is that the new silicon surface below the oxide has never been exposed to atmospheric contamination, so it should be as pure as the original material.

After the oxide has been formed, it is coated with a substance sensitive to ultraviolet light. This substance, called photo-resist, hardens and becomes insoluble in certain acids wherever the light falls. The next stage is to place the photomask, which defines the patterns of windows to be opened in the oxide layer, on the wafer and illuminate it with ultraviolet light. After the exposure, areas where the openings will be are still soluble, and the rest of the wafer is protected.

When the wafer is washed in acid, the soft photo-resist dissolves, leaving areas of exposed oxide. Further washing in another chemical will then dissolve the oxide but leave the exposed silicon surface and the hardened photo-resist unaffected. Finally, the remaining photo-resist is dissolved.

To recapitulate: The aim is to introduce phosphorus into certain precisely defined areas on the wafer below the silicon surface, leaving the rest of the wafer untouched. These areas of bare silicon are positioned using the photomask. All that remains is to put the wafer in a furnace, this time with a phosphorus atmosphere, and let the phosphorus work its way in. But how can a gas work its way into a solid? The answer is . . . very slowly.

Diffusion is the name given to the process whereby gases will mix and spread to fill any volume available to them. It has its origins in the random thermal jumping around that the molecules do by virtue of their temperature. Gases can mix in a few minutes or even seconds, but for a gas to penetrate a solid surface takes hours, even at the high temperature of the furnace.

Yet it is this very slowness that, as with the oxide growth, leads to the possibility of extreme precision in controlling the depth of these layers, which are, typically, only a few micrometers deep. By carefully regulating the temperature and the time, the depth of penetration of the phosphorus can be controlled to within a small fraction of a micrometer. The rate of penetration of the phosphorus into the oxide layer is so slow that it can be ignored, so the oxide has done its job of allowing only a selective introduction of phosphorus below the silicon surface. Any phosphorus caught in the oxide is electrically inert since it is surrounded by an excellent insulator.

We now have an array of buried microelectronic circuit components in the silicon wafer, protected by an insulating layer on top and absolutely untouched by human hands. All that remains is to provide them with metallic top contacts that can then be used for electrical connections to the outside world. These diodes could have diameters as small as desired, down to the limit of *resolution* of the optical system (that is, how close two lines can be without blurring together), which is a few micrometers, one or two micrometers at best.

The preferred metal for top contacts is aluminum, and this is deposited as a thin film by evaporation. The wafer is placed above a small crucible of aluminum in a vacuum chamber, and a film of aluminum about one micrometer thick is deposited. The aluminum can be selectively removed from the wafer, leaving only a set of contacts on the diodes, using another photomask and the same procedure as with the oxide layer. Again, selective chemical action by special solvents is the key to defining the pattern of top contacts.

This "circuit," as I have described it, is simple, but all the ingredients are there. In an integrated circuit such as a microprocessor, there could be several buried layers at different depths, produced by successive diffusions in atmospheres having different additives. There could also be more layers added to the original surface, perhaps including two metallic layers separated by an insulating layer and connecting with different parts of the transistor structures beneath. The more layers to a circuit, the more photomasks are used, and the more acute becomes the problem of vertical alignment between the layers.

Photolithography is at the very heart of microtechnology; it uses a stack of two-dimensional layouts, connected by vertical links, to build an impressive three-dimensional structure without using wires, soldered joints, or separate electrical components.

We are now almost ready to complete the process that began with a single crystal of silicon and will end with a packaged integrated circuit.

Each chip on the wafer is electrically probed by a computer-controlled circuit tester. Defective circuits are marked with a spot of ink, and will be discarded later. No attempts are made to repair circuits that do not work, since it would be impracticable and wholly uneconomic. Frequently as many as 85 percent of the circuits have to be rejected. If this sounds excessively high, consider all the critical stages involved. Perhaps we should be impressed by how many of them work rather than by how few.

The wafer is sectioned into individual chips by marking a grid of scratch lines and then breaking it. The working chips are mounted in protective packages with thin aluminum wires joining their contact pads to the pins on the package. Finally, the complete device is tested again; it may also be subjected to various vibration and temperature-cycling tests to ensure its reliability.

After the mass-production stage, the costs rise steeply because the chips then have to be handled individually and often manually. For this reason, there is constant effort toward methods of further automation.

Now that the basic processes of silicon chip production have been outlined, we need to look more critically at the limitations of present technology so that we can see where the next steps will lie.

THE BLURRING OF THE LINES:
A PROBLEM OF RESOLUTION

Consider diffusion first. This is the means whereby phosphorus is introduced through carefully defined openings in the oxide layer. Those openings defined the areas of the diodes beneath. Or did they? The assumption underlying the use of photomasks is that the patterns on the mask will be faithfully reproduced on the wafer. The packing density of the components is so high that any spreading effects under the windows may lead to unpredictable performance or even internal short-circuiting and consequent failure.

Diffusion, unfortunately in this case, does not cause the incoming phosphorus to keep straight when it has passed through the opening. A crowd of people jostling to get through a narrow door have the tendency when they get to the other side to spread out, to move from a region of high concentration to one of lower concentration. The phosphorus atoms move laterally under the oxide by about the same distance as they move vertically below it. This has two immediate consequences, one good and one bad.

The bad consequence is that the area of the diode formed is greater than the area defined for it by the photomask; and that will alter its electrical characteristics. The good consequence is that the interface between the phosphorus-impregnated silicon and the boron-impregnated silicon is protected by the oxide. This is an advantage because such junctions are particularly sensitive to contamination.

There is an alternative to diffusion that ensures that the phosphorus atoms continue in straight lines after passing through the window in the oxide. Perhaps the best analogy here is one of firing bullets at a metal plate with a large hole in it. The high speed of the projectiles guarantees straight-line paths; it would take a collision with a massive obstacle to deflect them much to one side after passing through the hole. So it is with this method of implanting the phosphorus into the silicon: The phosphorus atoms are electrically charged and then accelerated using high voltages (100,000 volts would be typical) toward the wafer; the whole process takes place in a vacuum chamber and in this case the temperature does not have to be high. The pattern of the diode formed in this way is an exact shadow of the window in the oxide, giving this method great precision. Penetration depths are limited, though, to about half a micrometer even with these high accelerating voltages, so this is a limited process. Also, the advantage of diffusion is now lost and the buried interface is no longer protected. As so often happens in an engineering design, a compromise is warranted. In any given application the relative merits of precision circuit definition versus a protected interface must be assessed.

The second problem to consider in looking at the limitations of conventional photolithography is the way high-precision stencils, the photomasks, are used in practice.

In the description of how the areas of hardened photo-resist are defined on the wafer, I mentioned pressing the photomask against the surface before turning on the ultraviolet light. Direct contact has the advantage of producing the sharpest shadow, but it also means that any dust particles on the photomask will be ground into the wafer surface, damaging both the wafer and the photomask and making the useful life of the photomask a short one. Often, to ensure the best possible contact, a vacuum is created between the two surfaces so that air pressure forces them together. It does not take many operations of this kind before the cumulative defects on the photomask make it unusable.

Could we perhaps support the photomask just clear of the

surface so that its shadow is projected onto the wafer without any physical contact? The life of the photomask would then be virtually unlimited. With the best optical systems available, this can be achieved without any loss of detail, but only over a restricted area of the wafer. It is a fiendishly difficult task to construct a lens capable of projecting an image with detail down to one micrometer on a region ten centimeters across.

Using the methods I have been describing, the very finest detail that can be produced is a line one micrometer thick—in principle. In practice, on the scale of production required to make the process economic, this limit is nearer three micrometers. The yield of successful circuits is typically 15 percent with this level of detail. If we were to aim for the theoretical limit of one micrometer and only accept those circuits that met that specification, the yield would probably drop a thousandfold and the price per chip would rise accordingly.

It looks as though the theoretical one-micrometer line has turned out to be the best we can draw. Is it good enough? In engineering research, even the best is never good enough. "The best" is simply our way of referring to the technology that we have to live with while we work on the next improvement. How to decide on the next step? By asking what it is that sets the one-micrometer barrier. When that is understood, it can be overcome.

The one-micrometer barrier is the consequence of using ultraviolet light. If red light were used, it would have been still worse. Have you ever wondered why ordinary microscopes never have magnifications much in excess of 1,000? Why aren't they made with magnifications of, say, 10,000? It could be done, but it would not be worth the effort because no more detail would be seen at a magnification of 10,000 than at 1,000. It is not the fault of the microscope, it is the fault of the light. The fact is that sharp edges do not cast sharp shadows: The light is scattered and the edge of a shadow has a slight blur on it. Two thin lines on a microscope slide will blur into one if their separation is comparable with the wavelength of the light being used. And that is the necessary clue: use short wavelengths. Blue light has a shorter wavelength than red, and ultraviolet is shorter still, but obviously not short enough. Obviously, light is just not good enough for the job. The next step must be taken in the dark—unless you happen to have X-ray eyes.

LIGHT IS TOO CLUMSY
In pursuit of ever finer lines on the microcircuit diagrams, we must look for illumination with shorter wavelengths than ultraviolet light, and that means X rays. They have wavelengths ranging from about one-hundredth of a micrometer right down to the gamma-ray threshold of about one-hundred-thousandth of a micrometer.

X-ray lithography uses exactly the same principles as those in photolithography. It is a comparatively recent addition to the available tools in microelectronics, since it took some time to find suitable chemicals analogous to photo-resist, which will harden on exposure to X rays. An X-ray mask is made from a pattern of thin metal strips on a thin membrane, and is rather fragile. The thin membrane is needed because otherwise the X rays will not get through. A factor limiting the use of this type of lithography at present is the scarcity of powerful X-ray sources. There is nothing yet to compare with the ultraviolet floodlight. This in turn means that the exposure time is long compared with that needed in conventional photolithography. Finally, the X rays do not get far in air before they are absorbed and scattered, so the exposure must take place in a vacuum or in an atmosphere of helium.

None of these difficulties has proved insuperable, and microcircuits with line spacings of less than one-tenth of a micrometer have been produced. Whether this form of lithography is developed fully will depend upon how another short-wavelength process fares: the technology of electron-beam lithography.

Do you regard electrons as particles or as waves? This is a loaded question, and one that caused many heated discussions in the early part of this century. When a dispute becomes unduly protracted, when the answers seem no nearer, it is time to examine the *question* to see whether this is right.

In this case, the question was wrong and the issue is now dead. Electrons, whatever they are, have some wavelike properties, some particlelike properties, and some other properties too. The dominant characteristic depends upon the nature of the experiment.

If you make a living designing cathode-ray tubes for television receivers, you will find it very convenient and very accurate to regard electrons as charged particles. If you are a crystallographer, finding out about how the atoms are arranged in solids, you will interpret the pictures you get (on a cathode-ray tube) in terms of electron waves and things called diffraction patterns. If you work with electron microscopes, you will see something of both worlds. The process of microcircuit printing using electron beams has benefited from several decades of earlier experience with scanning electron microscopes. The electron wavelengths in these instruments are so small that useful magnifications of several million are possible. They look like promising tools for our purpose.

With photolithography and also with X-ray lithography, the sensitized surface of the silicon wafer is flooded with ultraviolet light or X rays through a microscopic stencil called a mask. The pattern on the stencil is thereby transferred to the surface of the wafer, together with any imperfections the mask might have. One of the powerful and attractive features of the electron-beam approach is that the mask is unnecessary. The pattern of the circuit is written directly onto the electron-resist by a very narrow electron beam controlled directly from a computer. Where the tiny focused spot of the beam falls, the electron-resist hardens; thereafter the series of operations to open windows in the oxide, diffuse other substances into the silicon, and deposit metallic layers proceeds as with conventional photolithography.

The fact that the beam must scan every circuit pattern on every chip outline on every wafer, one at a time, means that the process takes much longer than either of the others considered, so there are some drawbacks. All in all, though, it is a distinct improvement. Layout details down to one-tenth of a micrometer can be reliably produced, and the circuit information that reaches the silicon surface comes directly from the memory of the computer.

Before we leave this brief sample of silicon technology, I will return to the crucial issue of the yield of successful circuits from a crystal. If it could be pushed up to 90 percent the costs would fall still further. One of the really difficult problems is that the crystal of silicon has a high density of defects when it is first formed, due primarily to the stresses involved (a single crystal may weigh ten kilograms). These defects take the form of dislocations in the otherwise perfectly regular atomic lattice.

These defects are distributed randomly in the crystal, and unfortunately their presence is not detected until a complete circuit is given its first electrical test. The yield of successful circuits is found to fall sharply as the size of the circuit increases, and the reason is not hard to find. The larger the area of a circuit, the more likely it is to include a defect, and it only takes one defect to cause total failure. In the extreme case of a circuit so complicated that it occupies the entire area of the wafer, a yield of one good circuit per thousand, perhaps even less than that, is reasonably expected.

These circuits will come one day, but can we expect to produce significantly better silicon by then?

ONCE MORE UNTO THE VACUUM . . . ?

Electronics began with a vacuum inside a glass bulb. What of the vacuum beyond the Earth's atmosphere? How about an ultra-clean orbital laboratory where the apparatus is open to the stars and the people doing the experiments carry their own atmosphere with them in space suits? The vacuum as such may not be worth the effort of an orbital laboratory, after all, we can produce adequate vacuums down here on Earth. But the other characteristic of an orbital laboratory is not so easy to duplicate: free fall, that is, very low gravity.

The growth of a truly perfect, massive single crystal of silicon might be feasible in orbit. It would solidify, perhaps over a period of weeks or months, as a near-perfect sphere, kept from touching the walls of the test chamber by electromagnetic levitation. It's a thought, anyway.

SUMMARY AND TECHNICAL TERMS

Silicon is one of a class of materials known as semiconductors; their electrical properties are intermediate between those of good conductors like copper and good insulators like plastic. Examples of other semiconductors include germanium (the material from which the first transistors were made) and gallium arsenide (used for lasers on a chip and for the light-emitting diodes—LEDs—in calculators and wristwatches).

Two basic types of transistor are used as electronic switches on silicon chips: the bipolar transistor, which regulates the current flow in a similar manner to water flowing over a variable-height barrier from a lake; and the field-effect transistor, which regulates the current flow by constricting the channel available, like a hose being squeezed to control water flow through it.

The silicon for use in integrated circuits must be of high purity and also it must be a single crystal. The purity is achieved by a process known as zone refining, in which a traveling molten zone sweeps a high proportion of the impurities to one end of the rod. The crystal-growing is carried out by dipping a perfect seed crystal into a crucible of pure, molten silicon and then gradually withdrawing it (the Czochralski process). It is at this stage that the vital trace elements (known as dopants) such as either boron or phosphorus are added to the molten silicon.

Silicon chip integrated circuits are designed using an interactive computer simulation. The designer can, in effect, experiment with different circuits until the required performance is predicted. An axiom of circuit design is to use the minimum area possible so that maximum density of circuits per silicon wafer can be achieved.

The key to the low cost per chip is that hundreds of silicon wafers, each containing hundreds of individual chips, can be processed simultaneously in true mass production.

The use of multiple layers above and below the original surface of the silicon wafer (planar technology) makes it possible to manufacture and isolate or interconnect all the transistors, diodes, resistors, capacitors, and other electrical features without the need for handling individual components. Each layer of the circuit has its own pattern defined in turn by a shadow-printing process called photolithography. In this, the image of the circuit is projected using ultraviolet light upon the sensitized wafer surface, causing chemical changes in the surface coating. When the wafer is immersed in the appropriate solvents, a pattern of windows is formed in the protective oxide layer through which further dopants can be absorbed into the silicon inside a high-temperature diffusion furnace. These diffusions give the integrated circuit its vertical dimension. The depths involved are all very precise because the layers are formed at such slow rates, typically one micrometer or less per hour.

Photolithography in routine production can resolve circuit details (line separations) down to a few micrometers. This limit is set by the wavelength of the ultraviolet light used. The lateral spreading of the dopants under the edges of the oxide windows means that the circuit pattern is not an exact copy of the photomask. This last difficulty can be overcome by blasting the dopants through the windows using a process called ion implantation. The circuit detail can be improved by a factor of at least ten if X rays or electron beams are used instead of ultraviolet light. Electron-beam lithography has the advantage that the beam writes the circuit pattern directly onto the sensitized silicon surface without the need for a photomask.

▪ 4 ▪
Beyond Silicon

Silicon techonology is the foundation upon which most of present-day microtechnology rests. Yet there are things that even silicon cannot do and never will be able to do. In this chapter I will describe some chips that use materials called garnet, or lithium niobate, or metals such as tantalum and niobium. There is no need to memorize these names, the main principles can be understood through simple analogies.

The first non-silicon chip I will tell you about is a memory chip that uses a substance called garnet, the same material that is used as a gem. Other chips use sapphire and diamond as well, but that is another story. Various sorts of memories, that is, information stories, are used in computers. What would be the specification for an ideal computer memory? It should be able to store an infinite quantity of information, any item of which should be available in zero time for the rest of the computer to use. And one other thing: There should be no risk of amnesia.

Strange though it sounds, computers sometimes use memories that could fail if the power supply to the computer were cut off. It can be very amusing to watch the expression on the service engineer's face when he opens the bottom panel of a million-dollar supercomputer to reveal . . . a car battery. The embarrassing truth is that without that standby power supply, all the information in at least part of the computer's memory might be lost. Naturally, there is great interest in ways of storing information more reliably.

Long-playing records and tape cassettes are both information stores of the sort we might use. Magnetic tape, in particular, is available in vast reels and its retention of information does not depend upon the presence of an electrical power supply. However, getting at the information can take seconds, even minutes, in other words, eternity. That may not matter for some applications, such as retrieving a list of addresses for standard letters. But it would be serious, perhaps fatal, in a lunar landing craft or in a battle computer.

The trick, then, is to arrive at a good compromise between the qualities I have mentioned. A very promising candidate in the field is a type of memory chip that has at least one property in common with an elephant.

BUBBLES NEVER FORGET

In recent years, computer memory chips made from garnet and able to store useful amounts of information have been developed. They utilize magnetic bubbles. Information from any part of a bubble memory can be retrieved in about one hundred microseconds, which is fast enough for many purposes (such as generating a television picture).

The important thing about the bubble memory is that it retains its information even if the power supply fails. The information, you will remember, is stored simply as a pattern of things that are either on or off, representing binary 1 or 0. An alternative to on and off is present or absent, which is the case with magnetic bubbles. The bubbles themselves are only a micrometer or so in size and getting a million onto a normal-sized chip is no problem.

As long as the pattern in the memory is preserved, the gems of wisdom remain intact and accessible. Suppose that you are calling the roll in a class of schoolchildren in two different sets of circumstances, in school and at a fairground, where you will be taking them for the day.

In school, the roll consists of a list of names, and against each name the child's seating position as so many rows along and so many rows back in the classroom. You can quickly check the presence or absence of any particular child (corresponding to one bit of information) by counting along and back to see whether he is at his desk. Now off you go to the fairground together.

At the fairground, you decide that you should check once again whether they are all present, but there is a problem: The whole class is on the carousel. Can you still check the register? Yes, but it will take longer because although you are using the same class list (in alphabetical order, say) you may have to wait for anything up to a complete revolution of the carousel before you can be sure whether or not a given child is on it.

The pattern of the information is stored in different ways in different sorts of computer memories as I have described. The first way (the classroom register) gives you faster access to the information, but this type of storage sometimes fades out when the power fails.

The second way (the carousel register) is slower in access but quite acceptable for many purposes, and magnetic-bubble memories are of this continuous recirculating kind. The bubbles (or not-bubbles, as the case may be) march around the chip and the information is extracted from the memory by something equivalent to a person standing beside the carousel. The important thing is that even if the carousel stops, the pattern is still there and will be available again once the power returns. Similarly, the bubbles in the memory chip are immovable because of a permanent magnet until the power returns and they can begin marching again.

There is another nice thing about magnetic bubbles, apart from their resemblance to miniature elephants: These particular chips are very simple to make and there are no buried layers, with the happy result that it is possible to reprocess those with low yields, rather than throw them away.

Magnetic-bubble memories are advancing very rapidly, but not quickly enough to remember, say, half a million printed pages complete with diagrams. For that sort of job (an electronic library) there is another kind of memory. . . .

SEEING IS RETRIEVING: OPTICAL MEMORIES
Next time you see a long-playing record try and think of it as a conveniently portable information store, as a memory, in fact, for that is precisely its function: to recall a piece of music, for example, with great precision and attention to detail. Cassette tapes are the same sort of thing, with the further advantage that in some cases they start with an open mind (that is, you buy a blank tape) that can then be filled with new material.

The optical memory looks like a very flexible and shiny long-playing record with a diameter of about thirty centimeters. It is made of thin metal foil sandwiched between transparent plastic layers. The foil is the recording medium, the information is stored as a pattern of punched holes along the tracks, and by now you will be quite familiar with the idea that the 1's and 0's are represented on the disk by either the presence or the absence of these punched holes. The holes are very small indeed, about one micrometer across, and when the recording is made they are produced, in rapid succession as the disk rotates, by burning out the metal with pulses from a miniature laser held in the pickup arm. The laser is made from gallium arsenide in a chip about one millimeter square and is quite harmless.

There is no reason at all why you should not use this sort of disk to record music or moving pictures, it is quite commonly done, in fact—but its most likely use in the immediate future is probably in the realm of document storage and retrieval, that is, a library or an office filing system. Such a disk can indeed record half a million printed pages including diagrams, and play them back to you on a television screen.

The playback works by shining a low-power light source (in the pickup arm) onto the spinning disk, focused through the plastic precisely on the metal foil. If a hole passes under the spot of light, there is very little reflection; if the metal is there, the reflection is high. So the 1's and the 0's are played back as either "high" or "low" levels of reflected light.

The thing that is so impressive about this type of optical long-playing record is that there is no physical contact between the recording medium and the pickup head, which means that both will last forever. As if that were not enough in itself, surface scratches and abrasions on the disk don't matter either: the fact that the beam is focused through the plastic onto the protected metal foil means that surface dust (microgrot) will only slightly affect the beam, certainly not enough to confuse a *1* with a *0*, which is all that matters.

This disk will store about as much information as a normal long-playing record (ten thousand million bits, give or take a zero) but not all long-playing records are yet in this form, and the type of record player you need is clearly a little out of the ordinary. But the time will come. Digital recording, as it is called, has a number of advantages over analogue recording, but again that is another story.

There is a second optical memory device called a holographic memory. Looking at a hologram is like looking through an open window. Things beyond the window appear in their correct perspective, and if you move your head from side to side you can see around them. How far you can see around them depends upon how close they were to the photographic plate when the hologram was made.

Another sort of hologram makes things look as though they are on your side of the window. Only when you reach out to touch them and your hand passes through the solid image do you realize this. Good holograms are startlingly effective, and one day televisions will be like that, although with much larger screens.

The essential point about the hologram is that it contains more information than an ordinary photograph because of the fact that you can see around and behind the objects when you look at it. Naturally, these memories are of interest for use in computers.

Holograms can be made on pieces of celluloid that look a little like ordinary photographic negatives, until examined closely, when they seem to be just nonsense. If you hold a holographic negative up to the light, it seems as though you are looking through the overlapping folds of net curtains: just irregular rings and fringes. To make it come to life it must be illuminated with a light source having a particular color or combination of colors and coming from a particular direction. You will see why the direction is so important in a moment.

To build up a large memory from a series of holograms, they can be strung together into a long tape like the frames on ordinary cinematic film and projected in the same way. It is more interesting, though, to arrange them so that they are stacked against one another like cards in a pack. This is called a thick hologram. Suppose that every card in the pack (that is, every hologram in the stack) requires a slightly different angle of illumination to make it come to life. We turn on the light and slowly rotate the stack. What do we see?

Well, that depends on what was recorded. Consider two possibilities. First, each card in the pack shows a page from a book. Not very interesting, you may think, but by rotating the pack we are turning the pages of the book, and they appear and disappear in sequence. Thousands of pages can be stored in a single crystal of lithium niobate not much larger than a sugar cube.

The second possibility is to store a sequence of holograms taken like successive frames in a film. Then, a three-dimensional moving image is seen (in full color, of course). This concept was science fiction only a few years ago.

But on to more serious matters. The final topic we come to concerns a revolution in computers that has already begun, and will culminate in the adaptive hypercomputer any time now.

TANKS FOR THE MEMORY:
ENTER THE SUPERCHIP
Suppose that you set a marble rolling around the inside of an otherwise empty goldfish bowl, and it just kept on rolling. Could that happen? It couldn't, but something that sounds almost as outrageous can happen to a circulating electrical current if the conditions are right. That is, once it starts circulating it will just go on and on. Normally, a current going around a loop of wire would gradually die away, since the electrical resistance of the wire to current flow generates heat and so the energy is lost.

Some metals, such as niobium and tantalum, suddenly lose all their resistance to the flow of electric current if they are cooled below a certain critical temperature, close to absolute zero (which is a little more than 273° Centigrade below freezing). When that happens, the metals, which we would normally describe as conductors, become superconductors and their resistance drops to zero. These metals can switch very quickly from a normal resistance to zero resistance, and that means the two conditions can be used to represent those binary digits again.

The switches can operate in something like ten picoseconds (one picosecond is one-millionth of a microsecond). That is about one hundred times faster than the nearest that the silicon opposition can manage. It is also about twenty times faster than gallium arsenide. Admittedly you need a tank of liquid helium in which to immerse the computer, its memory, and all the interconnections, but that can be arranged.

Light itself can only travel a distance of about three millimeters in that time; since the parts of the computer cannot communicate with one another faster than that, it follows that these supercomputers must be no larger than the marble we started with if the snail's pace of a light beam is not to cripple their performance. Superfast and supercool, these are the embryonic hypercomputers.

SUMMARY AND TECHNICAL TERMS

Computer memories must meet several requirements, including storage capacity, speed of information retrieval, and reliability in the event of power failure. Magnetic bubble memories, using garnet rather than silicon, are an attractive compromise. A memory that will not fail when the power goes off is called nonvolatile.

The type of memory searched in the way like a classroom roll is called random-access memory. The carousel register is called a serial-access memory.

Optical memories include disks and holographic memories. The holographic cube memory or thick hologram can store many pages of a document or many pictures of a scene.

Computers based on superconductors (Josephson junctions, after Brian Josephson of Cambridge University, who predicted the effect) are potentially about a hundred times as fast as any silicon-based machine at present. They are close to one of the limits I gave in Chapter 1 because their performance is limited by the speed of light.

Research into new materials will continue to follow the two major lines I have outlined, semiconductors and superconductors, with some interesting variations. Following are a few possibilities.

Semiconductors have been investigated in the past using plentiful natural substances (silicon is the prime example). Interest is shifting toward much rarer substances, such as gallium arsenide, and the extension of this idea means that we will soon be *designing new materials* to specification—materials unknown in nature. Such designs will begin from a catalog of ingredients, the hundred or so known elements. The transmutation of base metals into gold is a very unambitious exercise when measured against the powers of modern alchemists.

The search for superconductors will concentrate upon finding—or making—substances that can superconduct at body temperatures. There has already been speculation that the human brain contains such materials. Nature may still have much to teach us in this respect, and computers of the future may well be grown in tanks of nutrient rather than molded in the heat of diffusion furnace.

• 5 •
Chips at Work

In this chapter we begin to consider the delicate and contentious subject of working with intelligent machines. Our attitude toward intelligent machines is determined mainly by our age. Most adults have never thought of computers as anything other than threats; most children have never thought of them as anything other than pets.

H. G. Wells wrote a short story called "The Lord of the Dynamos" in which a man worshipped a machine and eventually sacrificed himself to it. Fear and worship often go together, and perhaps they both have their origins at least partly in ignorance. Are we in the same position? Let me try out a word on you: *automation*. Its effect upon you is likely to be wholly determined by your age. Do you fear that you will be taken over by machines? Made to become like a machine? Be displaced by a machine?

If I were an extraterrestrial observer looking for evidence of intelligent life on this planet, I might be forgiven for concluding that, at least in the industrialized societies, we humans do indeed fear and worship machines. Why else would we try so hard to make ourselves in their image? Think of the boring, repetitive, dangerous, or dirty jobs that degrade the people who do them. I don't have to spell out what those jobs are. We seem to be trying to beat the machines at their own game. Why? They can never be like us; why should we try to be like them?

How about mad computers, or just unreliable ones? Can we trust them? Everyone is delighted when a computer issues a domestic electricity bill of a trillion dollars. If computers are that stupid, they can't be much good, can they? Do I really need to remind you where the computers get their information in the first place? It does not need much imagination to guess that many, perhaps most, of the computer "howlers" of this sort are in fact the result of human malice—that is, deliberate sabotage of programs or of data. Some people will go much further than that to demonstrate their dislike or distrust of computers. I am thinking of the California sheriff who, infuriated by an outpouring of paper in response to a simple inquiry, pulled his gun and shot the computer dead.

Computers, which are only machines after all, can go wrong. If their decisions are going to make the difference between life and death, can we make them more reliable? Yes, easily. They are so compact that it is quite feasible to duplicate or triplicate the vital decision-making parts. Blind-landing systems for aircraft do this. Naval pilots who must be prepared to land on the heaving deck of an aircraft carrier at night under storm conditions will let the computers handle the last crucial phase for them. You will not find many pilots, alive, who would seriously take issue with this.

Reliability, in the situations I have just described, and in many others that you can think of, is vital. This leads me to an interesting comparison. Suppose that two committees are appointed to report independently on a matter requiring rational decision-making, decisions based on information rather than on something else. The first committee has a single member and that member is a computer. How can we increase the reliability of the decision? By increasing the size of the committee to two or three computers and ensuring that they do not interact. If their decisions are the same (whether we are designing an aircraft or simulating a chemical plant), it is very unlikely that all three happened to make the same mistake or have the same fault. If their decisions are not the same, then there is a fault somewhere, and since they are machines it can be located and put right.

. The second committee has also only one member and that member is a human being. How can we increase the reliability of the committee's decision? If we make the committee larger, its decision will be less reliable; so we have to make it smaller.

Let's look into the office worker's world to see how silicon-chip computers are helping people to avoid becoming machines.

AN ELECTRONIC WORD PROCESSOR

I typed the manuscript of this book on an ordinary electric typewriter and I kept making changes as I went along. The result was a mess of deletions and transpositions that had to be retyped in order to be acceptable to the publisher and the printer. That is very tedious and not in the least creative. Wouldn't it be nice to only need the rough copy and then have it typed without all the errors while you or your secretary could get on with the next job? Such draft-excluders exist and are called word-processors. They are a step toward what we need.

The word-processor is just a moderately intelligent type-writer. As the rough draft is produced, the words are recorded, together with all the errors. As long as the errors are clearly identified (by being overprinted by the typewriter itself) then it will play back the document, minus the errors, at the rate of a page or two every minute (much faster if you want it to). If you want whole paragraphs interchanged, added, or deleted, you can instruct it, via the keyboard, to do these things before it starts retyping the rough copy. If you are sufficiently confident that one redrafting will be enough, you could also tell it to keep the right-hand margin neat and straight as well.

It will not identify and correct spelling errors since it is relying on you to indicate the parts you do not want reproduced (by overprinting them). A more fundamental difficulty is that even if the typewriter were equipped with a dictionary, the most that it could decide with confidence is that it did not recognize the word you used. You could easily have just invented a new acronym and forgotten to add that to its vocabulary. There is no question, therefore, of the machine "telling" you that you are wrong, because it cannot be sure. For that, it needs a new order of intelligence altogether.

It can be rather unnerving to walk into an office and see what looks like an ordinary electric typewriter working away while its master or mistress gets on with something else. The effect is still greater if the office is empty.

It's nice to have a chat with the office secretaries, if they are not too busy. The typewriter is noisy, by which I simply mean that you can hear it working. Surely we can do something about that? No problem. In principle, all we do is leave the keyboard in the office and put the rest of the typewriter in another room, where no one will be disturbed by the noise. In practice, there is danger that if you can't see what you have just typed, you might lose the thread. Clearly, it is an advantage to be able to see what has been typed so that it can be altered or continued in the right sequence. How can we do that without bringing all that noise back? A television screen can be mounted above the keyboard in the office. The characters and words appear on it as we type them, but with a further advantage: All editing can be done on the spot , simply at the touch of a key. Or, retrospective editing is possible as with the word-processor. Furthermore, there is no need to make the typewriter in the other room start working at all until the final draft is settled using the keyboard and display screen in the office. That saves paper, too.

Suppose, next, that you need access to files for lists of addresses, extracts from earlier reports, contract documents, and things of that sort in order to include them in another document. One way is to use filing cabinets and then start copy-typing; another is to have the files in the memory of the computer to begin with, so that they can be combined and edited with an absolute minimum of tedious work. Optical memories (Chapter 4) are quite suitable for this. The storage disk, for example, would enable you to get to any one page out of several hundred thousand within less than one second. In other words, it would appear on your screen within that time after you had keyed in its identifying code or simply its title, provided only that the title is unique.

Communications between places of work have now improved to the point where a book can be written by two authors, on opposite sides of the Atlantic, without any meetings being necessary and without any paper being used until the final draft had been mutually agreed upon.

Another sort of office where computers can help is the design office. In Chapter 3, I described how the designer would interact with a powerful simulation program so that, together, they could design a microprocessor to be built into a silicon chip. I would like to reemphasize the point I made then, namely that neither the design nor the manufacture of a microprocessor can be accomplished by unaided human effort.

The design of ships and aircraft involves detailed drawings as well as mathematical analysis, and although it will come as no surprise to learn that the computer can deal with the mathematics, it is worth mentioning that it can do all the drawings as well. Or at least, it can build them up in a flexible manner that saves the design engineer a great deal of time, freeing his skills for the more intellectually demanding work. Basic shapes such as rectangles, lines, circles and triangles can be called up at the touch of a key; they can be magnified, reduced, rotated in three dimensions and deformed according to precise rules built into the computer. Part of the power of such interactive design is due to the speed and accuracy with which the machine can predict the result of changes, whether great or small. The allocation of work between the two represents the ideal human-machine combination: Analysis is the domain of the computer, and creative synthesis is the domain of the designer.

An architect can experiment with the design of a whole city, and then get the computer to show him moving pictures, from simulated head height, as he strolls in his imagination around the city he has created. The same considerations apply wherever a perspective drawing would be useful. Another example might be the aesthetic merits of a new car body design. Three-dimensional and in full color, naturally; full size as well if the cost is considered to be justified.

SAFETY: ROBOTS AT WORK

A computer that can interact physically as well as intellectually with the outside world is called a robot. Jobs like paint-spraying, welding, and many routine assembly operations are just the sort of things that robots are good at doing. What sort of picture does the word ''robot'' conjure up in your mind? A clanking monster preparing to zap you with its laser-beam eyes? I hope not.

Industrial robots are seldom even mobile. Often they are fixed in position beside a moving production line, spot-welding or doing some other highly repetitive job. A robot lathe, for example, looks very much like an ordinary lathe. The human supervisor has to be there, though, in case things go wrong. Robots at present are not particularly clever, but they are improving. A less-than-clever robot welder, for example, just assumes that the pieces to be welded are in position at the right time, it does not check that they are there. Whether they are there or not, it does the weld, or tries to. Thinking of such things, I must confess to a totally irrational inclination to feel sorry for them rather than afraid of them. They have no feedback, and everyone needs feedback in order to get on in life.

Feedback is a word that has crept into the language of ordinary conversation although it started life as a highly technical term. You have feedback if you continually modify your behavior as a result of experience, that is, as a result of information received. That sounds very much like just learning, doesn't it? However, feedback involves *using* what you have learned, not simply remembering it.

A robot welder with feedback would have some means of sensing the positions of the pieces it has to weld, and moving them or the welding torch as necessary to achieve proper contact. A sophisticated machine might also check the quality of the weld as it is being made, and report defective work to the human supervisor.

Robots are still in their infancy; their potential is barely recognized, let alone realized. Some developments are being prompted by their convenience or efficiency, and some by their dire necessity. Here are a few examples.

Let's begin with the intelligent elevator. I expect you've had your fair share of annoyance from elevators whose doors won't close for those irritating extra few seconds, even though it is obvious that everyone is safely inside and there are no more people about to enter. On a long journey, it is not amusing when the elevator stops for the benefit of (apparently) invisible people. Well, that kind of thing need not bother us anymore.

Infrared sensors can tell the on-board computer that no one is near the door; it will close swiftly—only to reopen if by any chance someone does make a dash for the door. Similarly, if inappropriate buttons have been pushed it will be easy to cancel them, and if a stop is made in error, the passengers will be able to send the elevator on its way again with an absolute minimum of delay. If more buttons are pushed than there are people in the car, then someone is playing games, as the synthetic voice will politely point out before canceling them all and asking you to begin again.

It need not end there, of course. There is no need for empty cars to travel between floors at the same speed as occupied ones; they can go much faster. Similarly, the intelligent elevator will take note of the traffic density and adjust its movements accordingly to minimize waiting times. Inevitably, there will be pictures on the walls of the car—moving pictures. Advertisements, information on local weather conditions (a floor-by-floor commentary as you ascend the Empire State Building?), and stock prices are possibilities. Finally, it is worth mentioning that elevators need not be confined to vertical movements: There is no reason at all why they should not convey you, in a series of horizontal and vertical transitions, from the parking lot to your hotel room, or just across the corridor. And they won't expect tips.

What about situations in which the stakes are higher? This is where intelligent machines are going to prove valuable allies indeed. Imagine a busy airport, the air lanes already congested and the computer-controlled traffic flow far beyond any human's capacity for comprehension. A tire bursts on landing and the plane skids to a standstill, ominous smoke pouring from one engine. At the moment the tire bursts, a countdown begins. How many seconds before the rescue crew can direct the first foam hose at the engine? One factor is the frailty of the human body

under high acceleration, not to mention the risk of their own
truck's tires bursting if they accelerate too hard. The first on the
scene, after an acceleration that would turn a man into a mess of
jelly, could be a robot vehicle (controlled via television sensors
on the truck carrying the human crew) which, after equally
violent deceleration, matches speed with the skidding aircraft
and opens up with the foam gun. Such robot vehicles could
travel a mile in a few seconds by using a principle known as
electromagnetic levitation, powered by a grid of underground
cables.

Underground activities, such as mining, can be fraught with
danger. There is really no need today for people to risk their
health and lives in such dangerous places. The proper place for
the miner is surely in a very special chair in the company's
control office on the surface, while the robot miner cuts away at
the rock face ten thousand feet below. What is the significance
of the chair? The answer is *telepresence* or the feeling of being
there, even when you are not. Marvin Minsky, who works at the
Artificial Intelligence Laboratory of the Massachussetts Insti-
tute of Technology, developed a chair equipped with arrays of
sensors that fit around the arms like sleeves; they also fit over the
hands like gloves. Pressure, heat, or vibration encountered by
the robot are transmitted back to the human miner as sensations
of touch or warmth, just as though he (or she) were cutting rock
or shifting rubble themselves. Equally important, movements of
the controller's hands are translated into massively powerful
operations on the part of the robot: a true muscle amplifier with
the power of a giant. The applications of telepresence is also
considered in Chapters 4 and 11. Meanwhile, if mines still don't
scare you, how about the thought of strolling around the Three
Mile Island nuclear plant immediately after the accident?

This pinpoints the need for developments in robotics as a
matter of absolute necessity: remotely actuated tools of great

precision are needed so that we can move in, *fast*, to inspect and repair damage at nuclear plants instead of being in the absurd position of having to wait until the danger to human personnel is "acceptable." There is no need for any human being *ever* to enter a nuclear plant, whether or not that plant is functioning correctly. This simple point, incidentally, solves a whole series of security problems at a single stroke.

Let me finish this very brief glimpse into a fascinating subject by again referring to the implications of telepresence. It means that before long it will be as easy to operate machinery from the next county, the next country, or the next continent as it is from the next bench.

HELLO, MR. CHIPS

Conversational computers are becoming more intricate, but the standards of elocution set by their science-fiction counterparts are so high that their qualities are not always appreciated. Human beings are very sensitive to intonation and the flow of natural speech, but if we ignore that for the moment and see what has been achieved, it turns out to be very promising.

There are two sides to a conversation: the recognition of spoken words, and their synthesis in the form of a reply or statement. At the lowest level of computer intelligence, robot machine-tools and some household gadgets can respond to simple verbal commands, provided that these commands are already established in their vocabulary; that they are easily distinguishable from other commands; and that they are spoken clearly. There is one aspect in particular that will have far-reaching implications, translation machines. These are computers that can recognize a voice in one language and provide an acceptable translation, simultaneously, in any other. I say "acceptable" because there are profound problems in the correct translation of colloquialisms, different meanings in different contexts, and so on, but even if the earliest versions were hardly better than a "tourist's guide" they would still be useful. But I have omitted their main attraction—they will be small enough to fit in your ear.

• 6 •
Chips at Home

Silicon chips are already much in evidence in many homes, and in this chapter we will take a closer look at chips in children's toys and games, in cars, in television receivers, and, most important of all, in personal computers.

The immediate impact, the cutting edge of the revolution, is already manifested in the toy stores. Much more lasting, though, and much more challenging in its long-term implications, is the personal computer. The reason for the ultimate dominance of the personal computer over other computers is that it will be able to do almost everything that they can do, and a great many things that they will never be able to do. It will be the extreme embodiment of the guidelines I suggested in the first chapter: its memory and its speed will approach infinity, while its size and its cost will approach zero. As for its intelligence, well, we shall see.

CHILD'S PLAY

In the introductory chapter, I gave two examples of attempted predictions that were wide of the mark because they were based upon an excessively narrow viewpoint, namely, the natural but erroneous assumption that advances in a particular subject are built up logically and sequentially from foundations laid in that same subject. Matters often progress like that, but then they're not called revolutions. The impact of silicon chips in the home is beginning in an orderly, predictable manner, well exemplified by the first generation of cheap microelectronic toys; we will consider them first.

Given a moderately intelligent tool, it is natural to first think of it simply as a way of doing familiar things in a new way. So, we find that the earliest silicon chip toys to become common-place in the home (not counting digital watches and calculators) are linked to such television display games as tennis, handball, hockey, baseball, and soccer. Notice that the main attraction here is the novelty of actually having some direct control over what happens on a television screen.

The next stage is more rules and perhaps more participants. The point to note here is that they are still versions of familiar games. Examples of this include chess, battle simulations, quizzes, and spelling games. The communication with the toy is at this stage still mainly via a keyboard or its equivalent (such as pushing pegs into holes), apart from a few voice-activated mobile toys. It is at this stage that we also see the emergence of the first teaching games, a development of enormous import-ance because it shows the child that learning can be fun or, more pointedly, that fun can include learning. The implications of this for education are considered in Chapter 8.

The third stage is a game that allows the child to be creative, as distinct from demonstrating mere versatility within a pre-scribed framework of rules.

Every advance demands more computing power and a gener-ally higher level of intelligence on the part of the computer. Much effort has been put into the design of games that allow the human participant to be creative, and it has been found that children are consistently more imaginative and more demanding than adults. It follows that the most powerful computers should be set aside for children to use.

A limited creativity is allowed by an extension of the second stage to, say, a model railroad system that is sufficiently complex to make detailed timetabling of the trains (to avoid crashes) an interesting and intellectually demanding task. Or, in a battle, if the child can decide the resources, firepower, and mobility of each side so it is not obvious at the outset who will win, again there is a limited creative outlet. The third stage is reached, though, when the computer will cooperate with the child in inventing his or her own games: The child makes the rules and the computer obeys them rather than the other way around.

At this level of interaction it is too much of a handicap for the user to have only keyboard communication with the computer. Pictures are needed as well (Chapter 3, the design of a microprocessor, gives an illustration of this high-level interaction). Since we do not want to alter the child, we must ensure that the computer is altered to respond in such a way that the child finds the experience rewarding and stimulating.

For a child, even words and pictures are not enough; sounds and gestures might be a better specification. A gesture can then include the drawing of a picture on the television screen by the child (using a special electronic pen) and the squeezing or pulling of a handle or lever, for example. In response to these words and gestures, the computer might generate pictures with colors and textures that do not correspond to any known paints, or sounds that do not correspond to any known musical instruments. Voice recognition as such is not necessary for this particular toy. Consistent tones and intensities of sound from the child will produce corresponding images or sounds from the computer. Together they can explore new worlds of their own creation, without the constraints of other peoples' preconceptions.

A barrier to the third stage at present, and also a barrier to the universal acceptance of computers, is the need to learn some sort of programming language in order to communicate with the machines. A programming language is a set of words and symbols that have precise meanings for the computer, but these meanings may be far from self-evident to the user. Generally the instructions have to be typed into the machine using its keyboard. All these languages place an uncomfortable and unnatural constraint on us, because they require us to be consistent and logical, in other words to behave like a machine. We cannot change, so the computers *must* learn to accept natural language and pictures if they are to be of maximum service to all of us.

CHIPS AROUND THE HOUSE

Silicon chips are making their presence felt around the house in a number of ways. Consider the family car first.

Fuel economy and the control of exhaust emissions are the subject of legislation in several countries, and the rules will become more strict as the oil supplies dwindle. It is no trouble at all to equip a car with a moderately intelligent computer that will monitor the readings from sensors in the engine and continuously adjust the operating conditions for maximum fuel economy—given the fact, of course, that it can't alter your driving habits. Nevertheless, it could remind you from time to time of how much fuel you could have saved if you had followed its advice.

On the more serious matter of safety, there is much that can be done and is already being done. A car that has a potentially dangerous fault can be prevented from starting altogether by its inboard computer. Similarly, voice recognition chips would prevent starting except by authorized persons, and not even then if they happened to be drunk. The incorporation of black-box recorders and radar sensors would not merely give better road information; they would reduce the accident rate by transforming our attitude toward driving. Every one of us would think twice about taking even minor risks if we knew, beyond the slightest shadow of a doubt, that in the event of an accident there would be an exact record of the car's speed over the last few seconds, the condition of the engine, the visibility, the road conditions, the view in every direction at bumper height, and the mental alertness of the driver (as judged by the electrical brain rhythms) at the time. The record of these facts need not span a time interval of more than thirty seconds. A continuous tape loop would give the system a memory going back only that far, and triggered to freeze on impact.

For an important and influential minority of drivers, driving would suddenly lose all its attractions, and for the rest of us, it would become much more of a pleasure.

On a lower intellectual level, there are sewing machines that remember hundreds of stitch patterns and will reproduce them faithfully, and cookers that will start different stages of the main course at different times so that everything is hot and ready at the right time. Then there are cameras that you simply plug into your color television receiver that display the photograph, and will also provide you with a permanent copy if you want one.

On a more practical note, a very modest domestic computer can do the following tasks for you: keep a record of the contents of the larder and update your shopping lists accordingly; remind you of appointments; keep all the household budgets and warn you when bills are due or funds are the wrong side of zero; remember all your favorite recipes; tune in or record your favorite radio and television programs; answer the telephone when you are out; answer the telephone when you *prefer* to be out; keep all the names, addresses, and telephone numbers you need regularly and look up others for you on command; read your gas and electricity meters for you and inform the appropriate authorities; and make excuses for you when you can't pay.

A further stage of usefulness for the domestic computer is in the management of the energy consumption for your house. Once you have indicated your preference for areas of warmth or coolness, this remarkable domestic pet will open vents and windows, control the humidity, activate the solar panels, and pay for itself in fuel savings very quickly.

Now, more about television. Flat, wall-sized screens and three-dimensional color will remain on the expensive side for a few years yet, so I will concentrate first on the pattern of television usage.

For most of us, television still means entertainment rather than information or education. We don't associate it with work; however, this will gradually change.

The first steps have already been taken: in the United Kingdom the Open University makes use of television; and several public information services are in operation. These give access, via a telephone line and a modified television receiver, to what is essentially a large and continuously updated filing system and reference library. Examples of the available information include details of railway timetables, recent legislation, sports events, commodity prices, entertainment, employment, books and references, advice (legal, medical, financial), and houses and gardens.

The system is very simple to use. When you first dial the system, a list of subject headings similar to those just given appears on the screen. For simplicity, let's assume that there are ten such general headings, each with its number. You are provided with a small keyboard with the television, and using this you key in the number of the subject you want. This produces a new picture with ten more headings, this time more detailed, and again you select one of these. It would not be difficult to make six such decisions in half a minute: you would then have selected one frame out of a million in thirty seconds, and of course you can continue further until there is no more detail available from the system.

If you find that you really need a permanent copy of the information for future reference or cannot spare the time to just sit there and absorb it, then clearly it would pay to make the television receiver slightly more intelligent or, at least, to improve its memory.

This means that you would record the information for later playback in the same way that you would record any other picture or television program. This brings me to the changes in viewing habits.

The television as a reference library is a concept that will not only appeal to children who have homework to do, or indeed anyone seeking documented information. Life at present is such that we are generally only free to use our televisions at home in the evenings (or at night, don't forget that the information services are available on a twenty-four-hour basis) and this could lead to family arguments.

However, disputes over what to watch are already on the decline. One solution to these arguments is to have two or more television sets in the house. More economically, several different screens could be supplied, from the same receiver, in different rooms. One screen or part of a larger screen would be reserved for urgent messages and letters from friends, which could be electronically "mailed" only a few seconds earlier from the other side of the world (see Chapter 9).

Another way to alleviate conflicts at viewing time is to pre-record television programs; then you could simply dial them and watch when most convenient. There will be notable exceptions, of course: News and sports events are obvious examples.

You are probably familiar with pictures that show different angles or facial expressions as you move past them. The same idea can be applied to television at home: Several people can watch the same screen but see different programs. The sound will be fed to each person from directional loudspeakers, servo-controlled and mounted unobtrusively around the walls. If someone wishes to read a book in the same room, he can sit in a chair that is carefully positioned to make the screen look blank and where there is, of course, no sound.

There is something special about a "live" television transmission that makes it uniquely appealing, especially if there is plenty of action. However, another sort of live picture may become popular soon. You can hang one of your television screens flat against a wall, and on that screen display a favorite picture from an art gallery; perhaps the gallery is in New York, perhaps Paris or Rome, it makes no difference. The rental for such a picture would be quite modest and unobtrusive: In the cashless society, your credit would be adjusted automatically by mutual agreement between your computer and that of the museum. The picture on the screen would be a live transmission from a camera permanently mounted in the gallery itself, and the quality would be superb. But that is only the beginning.

How about a picture of the Alps or the Grand Canyon? This would, of course, be supplied by a fixed camera, transmitting in full color and in three dimensions. Or, to really impress your friends, you could rent the ultimate live picture—the picture of the Earth transmitted from the Moon.

We know about wall-sized screens, but how about screen-sized walls? Let's think what this could mean. You could redecorate the room every week, just sitting in the armchair. All that it takes is the pressing of a few buttons on the control console, and the wall changes its texture and pattern according to your specifications. An electronic pad would permit all manner of graphic (not to say graffitic) experiments. The pattern could also be gradually changed throughout the day, its colors blending to complement the weather outside. The room could even be made responsive to your moods.

FOR YOUR EYES ONLY: THE PERSONAL COMPUTER

We now come to another computer jargon word: *dedicated*. A dedicated computer is one that exists for a single application only and is not shared among different users or different jobs. It is the dedicated personal computer that I am concerned with here.

A dedicated personal computer will expect to receive instructions using a simple programming language that it accepts in typed form via a keyboard. It will also have a screen on which you can draw, using an electronic pen, and where it will display its replies, questions, and statements. Despite these gross limitations, it can still be of some assistance. Here are a few examples.

First, it will take the place of all your files and correspondence at home. It will produce neatly typed summaries of anything you need to send to someone else (such as completed tax return forms or an insurance claim).

Second, it will attend to all the domestic reminders and other forms of assistance I outlined in the previous section. If you are studying, it will store and collate all your notes for you, together with the contents of as many textbooks or extracts that you may want. An architect will have all his plans immediately available, together with any other relevant information he needs. A lawyer will have details of every trial that has ever taken place, and all the legislation ever passed, if that is what he wants. He may prefer simply to have records of all the games of chess he has played with it, together with its advice to him on how to do better in the future.

Engineers can try out their new designs; doctors can keep all their patients' records with them all the time; teachers can review the material for the next class or lecture as well as keep all confidential records permanently available; composers can hear their music being played as they compose it; and the hobby enthusiast can build a unique body of information on his or her favorite subject.

The capacity of your computer to interrogate the information networks has far-reaching implications. Here is one interesting thought. Imagine the effect upon someone preparing a political speech of knowing that any reference to alleged facts could be checked within a matter of seconds by everyone listening.

Home computing enthusiasts are breaking fresh ground today. They are so keen that in many cases they do not just buy their computers; they build them. No one knows exactly what this sort of revolution might bring about or what it might be possible to achieve. One aspect of this new trend I find particularly pleasing: There seems at last the real prospect of the return of the gifted amateur to the front line of scientific inquiry.

You are probably worried about the extreme confidentiality of much of the information that the personal computer will store. Remember, though, that the information stored in this way will be more, not less, secure than it would be in a filing cabinet in an empty office. Furthermore, personal computers can be made very secure by the use of voice-prints and code sentences that change every time they are used. Fingerprints can also be used. Eventually, you will only be able to unseal it with a minute tissue sample that is enough to establish your unique genetic blueprint, and is much more difficult to forge.

At present, personal computers are clumsy and in a few instances they are not even conveniently portable; but this will surely change. As we shall see later, though, even being portable is not good enough.

Here are two important questions arising from widespread use of personal computers that we will have to face when we consider education and communications (Chapter 8 and 9).

What are the implications for us when it is no longer necessary to be able to do arithmetic, or to read, or to write either?

How do you assign homework for a child who owns a computer powerful enough to solve differential equations, draw maps, translate Russian, simulate the dynamics of a supernova explosion, and beat the world chess champion, every time, at his own game?

· 7 ·
In Sickness and in Health

In this chapter we will consider how silicon-chip computers and other microelectronic advances can help our hard-pressed medical services, how they can help the handicapped and the disabled, and how some unexpected people are coming to think of computers as pets, not threats.

It is worth emphasizing immediately that to take full advantage of all that the silicon chip can do for us, we need to have better long-distance communications between computers as well as between people. These improvements are gradually taking place. The increasing use of glass fibers (Chapter 9) rather than copper wires, for example, means that information can be sent thousands of times faster. In other words, hundreds of television channels can be sent down one glass fiber simultaneously (and, yes, it is possible to unscramble them all at the other end). Those television channels are not for entertainment; they will contain X-ray plates, medical records, accident details, and laboratory test results.

This means risking a loss of privacy, but we must weigh the possible loss of privacy against the possible loss of life. I would not mind someone looking into my medical records to find out about blood type and specific drug allergies if that someone were trying to save my life.

The risk equations are never simple; their application is immensely complicated by the need to respect the wishes of individuals even if those wishes conflict with medical judgments. In pursuit of better health, what more natural place to begin than in your own doctor's office? Might silicon chips and glass fibers be able to help things along?

Medical records are still kept, for the most part, in card-index files. The card index is a form of memory; its capacity is determined by things like the size of filing cabinets. The time of access to the information is only a matter of seconds . . .or is it? That depends upon a number of things.

Ordinarily, this system works well; but suppose that you, your doctor, and your medical records were *not* in the same room at the same time; would it still work as well? If someone is involved in an accident or suddenly taken ill, especially if he is not near his home when it happens, then we can see some distinct advantages in the use of high-speed computer and television links.

If the person involved needs urgent treatment (whether by drugs or by surgery) then immediate access to his medical records by the hospital doctor is vital. The wrong drug could kill him if he is allergic to it or if it interacts dangerously with one he is already taking.

Assuming that he can be identified or can identify himself, the next problem is to get at his medical records. Those records are probably in the doctor's office, which may well be empty if it is late evening or during the night. A telephone call will either reach the person's own doctor or someone who is on standby, and he could get to the records, but how long might it take? Too long, perhaps.

If the person cannot be identified, then the hospital staff are faced with the necessity of making crucial, irreversible decisions on limited information. The problem of personal identification is again tied up with privacy and the freedom of the individual: the freedom to remain anonymous in the intensive care unit, in this case. Fair enough. Having your name and address (both of which might change) tattooed on your skin is perhaps not an attractive thought. In fact, this is not necessary. We could be safely identified by a skin marking that is invisible in normal light but will show up (in darkness, if necessary) when illuminated from a special lamp, such as ultraviolet.

How might this chain of events be altered by the existence of medical records in the memory of a computer? Simple: provided that the person in need of treatment can be identified, the records can be obtained from the standby computer in the doctor's office at any time of day or night, within a few seconds.

A moderately intelligent computer would first need proof that the inquiry is legitimate (a code sequence from the hospital would probably satisfy it) and then it would ask about the nature of the information needed. In an emergency, it would supply the vital material first, such as your blood type.

In this way, the communication system can, in effect, bring you and your medical records together. The third ingredient, your doctor, may not be physically present, but he or she could still be very valuable at the other end of a television link (based at his home or wherever he happened to be at that time). All these things can come to pass when the telephone lines use glass wires rather than copper ones.

This idea, that the doctor can in many cases be of assistance without being physically present at an interview, has been taken up and extended already. At Logan Airport in Boston, there is a link of the kind that I have described. It goes to Massachusetts General Hospital. Anyone who is sick at the airport is taken to a medical diagnosis studio where there is a television screen and a camera; there is also a nurse in attendance.

He is interviewed by a doctor who is in the hospital several miles away. The camera is trained on him by means of a control lever operated by the doctor. There is a couch with another camera and facilities for microscopic examination of blood and urine samples, as well as X-ray and electrocardiogram tests. All these results are assessed directly by the doctor at the hospital as soon as they are available; the microscope can be linked directly to a camera in the test laboratory.

Thousands of people every year use this facility. In some cases, the physical presence of the doctor is necessary; however, a system such as this helps to ensure that the best possible use is made of the limited medical resources available. It is a particularly attractive scheme for serving remote communities or in difficult weather conditions, since it enables the doctor to decide quickly whether his presence is needed. It could significantly reduce the risk of the doctor being unavailable for an urgent call while out on a routine one.

In the future there will also be *remote surgery*. We have the technology to enable the world's best surgeons to carry out delicate operations anywhere on Earth without leaving their consulting rooms. The telepresence devices (Chapter 5) can be refined downward as well as upward in scale; microsurgery is already here, and the holographic techniques have existed for a long while. The communications room, not the operating theater, will be the nerve center of the hospital of the future. Until we do have remote surgery, it will still be possible for expert surgeons to guide less experienced ones through delicate emergency operations that would otherwise be beyond their competence.

Silicon chips and their descendants are of great service to the handicapped and disabled. For the totally deaf, it is possible to implant electrodes in the inner ear so that electrical signals corresponding to outside sounds can be used to stimulate the part of the brain concerned with hearing. A medical team in Melbourne is exploring this with patients who previously had normal hearing. Their brains are already used to interpreting sounds, so this is entirely feasible. If this is successful, it might be possible to help people who are congenitally deaf.

Teaching deaf children to speak is another problem that is being tackled with the help of microprocessors. Work at the engineering department of Cambridge University has been concerned with "showing" deaf children their voices on a television display and comparing the sounds they make with those made by normal children. Their voices produce a wavy line on the screen; the exact shape of the line depends upon the sound of the child's voice. By altering his voice to match another fixed wavy line, the child gradually succeeds in imitating the normal voice. This is a good example of feedback (Chapter 5). The child compares the shape of his voice with the shape of the target wave, notes the differences, and alters his voice accordingly. He learns and then modifies his behavior using what he has learned.

People who need wheelchairs can be helped by a voice-recognition chip that controls a mechanism. The mechanism could be the propulsion system of the chair, it could be a door, a cooker, a television, or a telephone. A spoken command will activate it. Much has been written elsewhere about fully automatic and computerized houses: perhaps more should be written about who should be the first to benefit from them.

For old people who live alone, the greatest danger is that they will not be able to get help when they need it. A communication system in which a small radio transmitter is worn by the old person and operated by a simple switch in the event of an emergency may be one answer to this problem. Another possiblity might be to install an intelligent personal computer in the house or apartment. It could inquire as to their well-being at random intervals (regular intervals might be too irritating and machinelike). It would perform all the domestic functions mentioned in Chapter 6, and its voice could be whatever the old person wanted to hear: perhaps a favorite television personality or a child's voice. The voice could also change according to circumstances.

Would the computer panic and send for help if there were no response to its polite inquiries? Perhaps its owner is simply asleep, or perhaps he has gone out. An intelligent computer would have no difficulty at all in detecting and recognizing breathing sounds and also in judging whether there is anything abnormal in the respiration rate. Similarly it would be able to tell whether its owner had gone out. If out, it would be ready with suitable greetings upon the return, and if out for an abnormally long time, it would summon a search party.

Blind people should eventually be able to benefit from a form of artificial sight that is similar in its approach to the scheme for helping deaf people. It uses an electronic retina at the back of a lifelike glass eye and sends the electrical signals directly to the visual cortex, that is, to the part of the brain concerned with sight. Although we do not yet know exactly in what pattern we have to connect all these electrodes, it does not necessarily matter, which at first seems rather surprising. The point is that the person will soon learn to associate the pattern of bright lights they "see" with whatever pattern of illumination is placed in front of them. They will be able to recognize and avoid obstacles; with practice they will be able to read signs, too.

Recognizing printed words is easier than recognizing spoken ones because there is less variation in form. Computers are getting very good at this. Computers that read to the blind are now available, though their voices leave something to be desired in many cases. They will improve rapidly, though, and before long it will be possible to have books read by a very convincing imitation of anyone at all, or by a voice tailored to someone's individual preferences. Naturally, this extends to reading different characters' parts in different voices, and to the reading of plays in such a way that the play appears to be performed by an entire cast of actors.

I don't think anyone really believes that this sort of computer, when it becomes available, will only be of interest to the blind. It is worth recollecting at this point that the typewriter itself was conceived originally to help the blind and with no thought of any wider applications.

The miniaturization of intelligent computers has important long-term implications for medical practice. Consider first the problem of maintaining the correct sugar content in the blood of people who suffer from diabetes. There are compact instruments now available that look like pocket calculators that enable the diabetic to measure the sugar level accurately for themselves at home. A tiny amount of blood is analysed by the instrument. With the result of the test conveniently and rapidly available, the diabetic can then (if necessary) take steps to correct the level.

The next step will be the incorporation of an insulin source in the monitoring device, perhaps as surgical implants, so that the regulation process takes place continuously and automatically without the diabetic having to think about it at all. Warnings would be provided if the computer found that it could not control the levels or if the insulin needed replenishing.

As a second speculative example we might consider the problem of someone who knows that he is likely to have a heart attack. He carries appropriate drugs with him to take when he feels an attack to be imminent. How might an intelligent microcomputer on a chip be of assistance here? It could continuously monitor the heart condition, giving a much earlier warning of possible trouble, and it could control the release of the drug into the bloodstream so that the potential heart attack victim could live as normally as possible, unaware even that the drug had been administered.

At a further level of miniaturization, there are glimmerings of hope for truly universal preventive medicine, as distinct from the palliative medicine just considered. A body monitoring capsule small enough to be carried in the bloodstream could detect and report abnormalities at the cellular level. It could provide a very early warning system, perhaps even for cancer. These monitor chips might pass through the body before imparting their test results; or they might take up more or less permanent residence in various vital organs, acting as sentinels, always vigilant, ready to report any departure from normal body functions.

After early warning, the next stage is implanted microcomputers equipped with biological weapons with which to attack as well as detect abnormal cells. Even this will seem a feeble effort once the two revolutions in microelectronics and microgenetics have been combined.

Biofeedback. Heard of it? Here is an example of how it works. Show someone a meter that reads his blood pressure, and ask him to make the blood pressure go down just by thinking about it, while watching the meter. He experiments with different states of mind; some make the pressure go up, others make it go down. Eventually, by experimenting and concentrating, he can make it go down. With more practice, he can manage without the meter, that is, he can reduce his blood pressure by adopting and maintaining a particular state of mind.

It is interesting to monitor brain rhythms while this is going on. These electrical rhythms are generated by the activity of the brain itself and are far from being completely understood. They can be detected by painlessly placing pickup pads against various parts of the skull. It has been found that the pattern of this activity is broadly related to one's state of mind. Sleep has its characteristic pattern, as does concentrated mental activity or relaxed reading. These electrical waves can be displayed on a screen either in the form of a simple television picture or in three-dimensional (holographic) form. In the latter form, the impression is of rolling waves, vortices, and swirling shapes of such variety that there would seem to be a new potential art form waiting for those who care to exploit it. Could biofeedback be harnessed and controlled without conscious disciplined effort, using an intelligent computer? If it could, would it be of any use?

It is an established fact that flickering lights and certain low frequency sounds can cause people great distress. If the light flickers at around ten flashes per second, it can generate via the eye an electrical brain rhythm that interacts with the natural rhythm and can cause drowsiness, nausea, or even epileptic fits in otherwise normal people. Sounds, or rather infrasound, because it is felt rather than heard, can produce the same effects. So far, all that has been established is that alarming states of mind can be induced by ''pumping'' the brain rhythms via either the eyes or the ears. Now suppose that it is possible to find a pattern that induces other effects, such as light sleep, deep sleep, alertness, concentration on intellectual tasks, or relaxation. Such a pattern could be impressed upon the brain using the same techniques as those I described earlier for helping the deaf to hear and the blind to see. An unobtrusive implanted chip, perhaps under the skin behind one ear, could do the job. It would have to monitor the brain rhythms carefully and ensure that they were kept within safe limits—a very tight feedback control indeed, but without conscious effort on the part of the person concerned. All he would have to do is instruct the computer to lock into a specified mode, say the alertness mode. Who might find this useful? How about airline pilots, troops on active duty, surgeons, athletes, and examination candidates?

The imminent prospect of this degree of brain control leads me to mention my lifelong objection to existing alarm clocks. It is this: they only do half their job, because they only decide when I shall wake up. An alarm clock could surely be set to decide when I go to sleep as well.

Would you mind being interviewed by a computer rather than by a doctor? Thousands of people have experienced this, and their responses to this apparently inhuman treatment are very instructive.

Take the Logan Airport remote-diagnosis clinic. Not only do the nursing staff report that the patients like it, but some of the patients have even made a point of saying that they prefer these television visits to the doctor's physical presence. Is this hard to understand? It becomes less so when we remember that many people find it difficult to communicate with strangers at the best of times, and doubly so about something as personal as illness.

The Logan experience turns out to be universal. For many years, hospitals in London, Edinburgh, Glasgow, and other cities have been operating some form of preliminary screening by computer interview prior to consultations with doctors. This has been made imperative by the fact that there is not sufficient medical staff available to deal with every medical inquiry as it comes in. There must therefore be a selection, and that selection must be based upon need. The use of computer interviews speeds up the selection and frees the doctors for duties where their presence is more urgently required.

The computer interview in its most basic form consists of a series of questions that either appear on a television screen or come from a loudspeaker, accompanied perhaps by a video recording of a doctor. The patient responds by pressing one of three keys: *yes*, *no*, or *?*, the last of these signifying that they do not understand the question. These interviews can be conducted in any one of several languages. This is all rather unnatural, which makes it still more significant that people respond so favorably and so consistently to such procedures.

At the next level, the interview is wholly conversational, that is, questions and answers are spoken, though the answers must still be of the yes/no form. This is fairly common practice now. It is a convenient way of establishing the background medical history of the patient, and getting an outline of his particular symptoms. The interviews are arranged under specific headings (for example, stomach complaints) and the questions are correspondingly restricted in range.

A further level of sophistication places the responsibility for a preliminary diagnosis upon the computer, and this is also becoming common practice, especially in the United States. The results are impressive. The computer diagnosis correlates very well with the combined expert knowledge of the consultant staff, which is as it should be, since the diagnostic programs are compiled from precisely this source. Medical diagnosis is well suited to computers because it is highly systematic and draws upon a well-established body of objective knowledge.

Medical programs are also used to teach medical students by simulating emergencies, in which the students have to make decisions from a selected range of possibilities in a limited time. By introducing an element of urgency and realism into the medical training, these programs help future doctors to make correct decisions quickly. The effectiveness of these programs has been established beyond all doubt.

Computers are being used to store the knowledge and accumulated experience of the world's best surgeons, so that their skills are permanently available to later generations. Using television links, this expert knowledge can be made available to less skilled surgeons, enabling them to carry out emergency operations that might normally be beyond their professional competence.

Before closing this chapter, let us return for a while to the most important factors in this whole business: the patients and their attitude toward the computer interviews. The computer is very carefully programmed by the hospital staff concerned; it has been found that patients appreciate friendly, even chatty responses that help to put them at their ease. For example, if the computer learns that the pains have diminished lately, it expresses pleasure; if they have increased, it says that it is sorry to hear that. Above all, it is always extremely polite, a point that is evidently of great importance.

One particularly interesting result is reported from a hospital group concerned with general psychological testing using computers. They found that many old people were reluctant to leave the computer. On being asked why, it emerged that nobody had taken that much interest in them for years.

• 8 •
Chips at School

The implications of the silicon chip revolution for education are, in the long term, even more far-reaching than in any of the other areas of human activity. The underlying reason for this has been given earlier in Chapter 2: The awakening of computer intelligence means that we have at last made a tool capable of designing its own successor. Unlike human generations, all the lessons learned from one computer generation are used to improve the next; further, computer generations are measured in time spans that are very short compared with a human lifetime. These two facts will combine to produce an intellectual explosion at any time now. When the adaptive hypercomputer develops itself we will face the supreme challenge, a challenge that everyone has assumed will either come from outer space or will not come at all: the challenge of finding ourselves in the presence of intellectually superior beings.

FROM ENTERTAINMENT TO EDUCATION
Few people will respond with much enthusiasm to the question "May I educate you?" but if we ask, "May I entertain you?"

then it is a different matter altogether. The trick, therefore, is to make education entertaining. Radio and television were thought of primarily as media for entertainment when they first became available, and many of us still regard them in those terms. The Open University, school broadcasts, and study programs are all combining to add the extra dimension of education without too much of an additional financial burden upon the users.

Cost is the present barrier: Very few people are prepared to spend their money on a home computer that will only teach them mathematics. However, if it will play games, help them with their hobby, and run the household budget as well as teach them mathematics, they might reconsider. The leading edge of the revolution is teaching us through the toy stores and hobby centers rather than directly from the schools and universities.

At the lowest level of computer intelligence, there are teaching programs that are barely adaptive, using very little feedback. These programs store a series of step-by-step explanations intended to take the pupil up to a certain level of problem-solving proficiency in, say, elementary mathematics. At the end of each section are problems the pupil must solve in order to progress to the next section. The pupil usually sits at a keyboard and watches a screen on which the teaching material appears, together with the questions and the answers he types in. Everything is fine until he gets a problem wrong, which is where the difficulties may start. The program can only react in a way characteristic of the very worst type of teaching: It can only repeat what it said before and it can only offer the same problems again and again. It will ''mark'' the tests in the sense that it tells the pupil whether his answers were right or wrong; it cannot identify individual points of misunderstanding and clarify them. If the computer has been blessed with a little imagination, it will eventually suggest that the pupil consult the teacher. The program is not adaptive because it is not sensitive to the particular difficulties of particular pupils. It does not modify its behavior in the light of experience, which is what we mean by feedback.

I mention this rather depressing case so that we can appreciate more clearly the vast difference between these early, crude versions and the very best teaching computers available now. Why bother with computer teaching at all? For the same reason that we are obliged to use computers in our medical services and training: There are just not enough teachers available to give everyone all the help they need at the time they need it. It is a matter of necessity, not preference.

Before moving on to consider being taught by a computer, it is worthwhile spending a little time thinking about an alternative way of making the most of the limited number of teachers available. This involves utilizing the improved communications facilities now available, especially television links.

Teaching using television links is an attempt to reach a viable compromise between two considerations. First, the fact that individual tuition, if it is of first-rate standard, is the best method of all. Secondly, the fact that the ratio of potential students to available teachers makes group teaching a practical necessity in the present educational system.

Using such a television system, a teacher in a special studio, which may be a room at a university or in a research laboratory, addresses an audience of perhaps a few dozen or a few hundred people, scattered in different locations many miles away. Some of the audience may be in a lecture theater: others may be in small groups around their television sets at home.

The teacher has several cameras under his direct control. One shows his own face as it appears on their screens; another is aimed downward at documents or drawings on his desk; and a third might be surveying the audience in the lecture theater, equipped with a zoom lens so that he can see individual questioners during discussions.

In one form of this system, members of the audience can

address questions to the teacher so that everyone else can hear both the question and the reply. This has the advantage that the teacher can if necessary modify what he says next or revise earlier material, in short he adapts to the response of members of the audience. It is disadvantageous because by dealing with an individual question at length, he may waste the time of others who already understand that part or who may want help with earlier material.

A second form of this approach largely dispenses with these problems. Every member of the audience has an individual communication link to the teacher that can be used during periods set aside for personal tuition. The teacher might first cover some basic subject matter (such as the relationship between current, voltage, and electrical resistance in a circuit) and then set some questions for the audience to tackle on their own. If someone needed assistance with one of the problems, he would press a key, and this would register on a control panel in the teacher's room and he could then open the personal channel to deal with the students' difficulties. In some systems they could see him but he might only be able to hear them; in others they would be able to see and hear one another. If several people wanted help at the same time, they could either wait for attention or, if two or three had the same difficulties, a small number of personal channels could be opened for detailed discussion. This form of teacher/student interaction is often used in language laboratories, where everyone is in the same large room. But it does not matter where they are, they could be scattered over an entire continent.

How quickly this form of teaching evolves will depend, among other things, upon how rapidly we replace copper wires by glass ones (optical fibers) in our telephone and telecommunications links. Another factor is the educational impact of the personal computer.

AT THE INTERFACE

In order to understand what the personal computer can do for teaching, we need to spend a little while thinking about teaching itself.

I am not a qualified teacher, but I consider myself a qualified learner, having sampled the British educational system at every available level as a student. I try to teach as I would wish to be taught, which is as follows. The first priority in individual teaching is to establish an informal relationship between teacher and student, in order to allow a fully adaptive response by the teacher to the needs of the student. In other words, there must be no inhibitions on the part of the student about speaking up when he does not understand. If he is intimidated even once by the teacher, then the relationship can be poisoned from the outset. Learning may still take place, but with an unnecessary element of stress attached, and extra stress is not part of my prescription for helping students pass difficult examinations.

Once the informality has been established, it is possible to adapt the teaching to be of maximum assistance to the student. If a step in the sequence of ideas is too large, then it can be explained in smaller steps with more attention to detail. If there are still difficulties, a completely different approach to the problem can be tried. Eventually, the conceptual gap will be bridged, although it will take time and effort on the part of both concerned. I am not suggesting that learning is ever easy, but some ways are a lot less painful than others.

Computer teaching has been tried with very young children, with postgraduate students, and at most levels in between. Much, perhaps most, computer teaching leaves a great deal to be desired, but the eventual possibilities must be judged by the best, not the worst, that can be done at present.

The best teaching computer programs are fully adaptive in the sense that I have just described, and they make a point of giving praise at every stage of understanding. They are always polite and their patience is genuinely infinite, which is more than any human teacher could manage. They are often encouraging, especially when the student is clearly having difficulties. How does the computer know when this is happening?

The computer monitors the student's level of understanding by providing continuous assessment, that is, a stream of problems to be solved by the student. The difficulty of the next question it sets is determined by the standard of the student's answer to the previous question. Communication with the computer is at present mainly by keyboard and by drawing on a screen. The screen will always be useful; how long the keyboard survives in its present form will depend upon the speed of progress in voice-recognition computers.

The computer derives its basic material from its human programmers, that is, from the teachers themselves. Although the main sequence of its presentation is unlikely to vary, no one can predict in detail how it will teach, since it will adapt to individual students, no two of whom will be alike. A good program might take one thousand hours of work for the production of one hour of fully adaptive tuition. The work will have been worthwhile, though, if that program can then be made available to thousands or millions of people on an individual basis. The production of such programs requires the very highest professional teaching skills, and these necessarily include an appreciation of the psychology of learning.

Computers are good at teaching subjects in which there is a large body of factual knowledge involved or where logical or mathematical analysis is needed: subjects such as spelling, reading, languages, arithmetic, geography, and elementary science or engineering. They are not, and perhaps never will be, much help at subjects where facts are scarce but opinions are important, subjects such as ethics, literary criticism, painting, and sculpture.

To be widely accepted and genuinely useful to all of us, the computers must learn to understand ordinary language, sound, and gestures, and to respond with spoken words and moving pictures. This is especially important for babies and very young children, whose natural creativity must not be curbed if they are to obtain the maximum benefit from their new toys. Computers, I need hardly say, do not have to be box-shaped. They can look like teddy bears, plants, marbles, puddings, armchairs, miniature elephants, or, if you prefer, you need not be able to see them at all. The one thing you cannot do is ignore them.

Regarding young children, computer teaching does not, repeat not, mean anything so trivial as just doing what we have always done but doing it more quickly and with better pictures; it means a complete reappraisal of the aims of education.

An experimental teaching computer in the United States uses a novel approach to instilling the ideas of geometry into young children (it is too advanced for most adults). It is a robot turtle that will move around in any geometrical pattern, provided that you can define the pattern for it. The task for the children is to invent more elaborate patterns; their reward is to see the turtle perform for them. This does not of itself impress them with the intellectual challenge of Euclidean geometry, but it might be a first step toward avoiding the lifelong hatred of the subject that most of us acquire after being taught it at school.

There are many ways to make subjects more interesting to those meeting them for the first time. Learning to dance is a familiar example. Ballroom dancing could be ''learned'' from a study of footstep patterns on sheets of paper (believe it or not, this is how some people used to learn). Another way of learning it is by dancing with someone. Using the second method, you learn how to dance and you also learn some other things as well, usually pleasant ones; in other words, it's fun. We are back to something we noted earlier, namely that a key ingredient of successful education is to make it entertaining as well.

Here is a glimpse of how a powerful simulation game can

engage the interest of a child and test the child's ability as well. Perhaps the subject of the lesson is first aid. Cartoon characters from a favorite story book have a variety of misadventures involving minor accidents. After each accident, the computer seeks advice on what to do next by offering several possible courses of action that might help the injured person. The child makes a decision and is then given the consequences of that particular treatment. If the "patient" gets worse instead of better, the child will try harder the next time because of the sense of involvement with the cartoon character. At a slightly more advanced level, the child is told that it is the captain of a space cruiser about to embark upon a dangerous mission with a cargo of valuable supplies. Naturally, things start to go wrong soon after takeoff, and different members of the crew seek advice at different stages. Depending upon the advice they receive, the adventure takes a different course every time the game is played; the final outcome is always the logical consequence of decisions made by the child.

It is not difficult to see how simulation games of this sort can be made to test knowledge and judgment in many different areas and in increasingly realistic situations. The sense of involvement, excitement, and responsibility will hold the child's attention in a way that combines education with entertainment.

Teaching one subject at a time is the rule rather than the exception at present. This will change as the need for expert knowledge in specialized fields diminishes, and computers store all the details for us. People whose training has in the past involved the memorizing of large amounts of detail, such as doctors, engineers, and lawyers, will spend much more time on fundamental principles and on considering the capabilities and limitations of the traditional fields of study in tackling realistic problems. Education will require less memory but more thought.

Realistic problems are always tackled by using a combination of the subjects we have traditionally taught separately. Here is one example: In deciding upon how best to conserve limited mineral resources, it is necessary to take into account geology, ecology, economics, engineering, and many other subjects as well. To tackle such an issue intelligently needs an integrated approach; here I am using the word "integrated" in much the same sense that distinguishes an integrated circuit from the old sort of circuit in which electrical components were made one at a time and then connected using wires and solder. I simply mean that all the aspects of the problem are considered together, rather than the question being considered from each viewpoint in turn or, even worse, from one viewpoint only.

In the past, only a highly educated person would be capable of taking this integrated approach, because so few people have a thorough understanding of the component subjects. Once the computers become the repositories of expert knowledge it will only be necessary for us thoroughly to understand the capabilities and the limitations of all these traditional fields of knowledge, so that we can formulate the problems in such a way that the computers can work out the details for us.

The source of this expert knowledge, the personal computer, can be regularly updated from a central computer using television links, and the central computer would in turn be updated from a selected few world authorities in the fields concerned. Yes, there will still be people who devote their lives single-mindedly to one field of study, and their contribution to our culture is of inestimable value. My point is simply that such experts should *never* be allowed to make far-reaching and irreversible decisions that require understanding of subjects beyond their own specialization.

I have stressed education, but training is important as well. When moderately intelligent television receivers are common, with built-in computers able to interrogate the central computer for us, everyone will have access to complete reference libraries that are regularly brought up to date and contain a multitude of training programs. There is a large measure of overlap between training and education, but we should not forget an important difference of emphasis. Training is for jobs; education is for life.

EXIT THE THREE R's

The three R's are reading, writing, and arithmetic. When computers can recognize print and read it to us, when they can recognize speech and turn it into print for us, when they can do all our sums for us, what will we do? When children have computers capable of playing chess at world champion standard and able to do all the mathematical analysis that is normally expected of graduate professional engineers, what are the implications for education? What will there be left to do?

Well, for a start, and only a start, we will be free to begin work on the really important issues facing the human race. These problems would today be rated in the "impossible" class. Here are some examples: a balanced world economy in which no one is starved of food or of education, control of world climate and weather to make the deserts bloom, understanding ourselves, making the best possible use of our limited mineral resources on a global scale, having a theory of the rainbow that includes the poetry as well as the physics, formulating the first thousand-year plan for the Earth while it is still beautiful and before it is too late.

Does all this sound a bit much for a five-year-old to start thinking about for a school project? I mentioned in Chapter 6 that the most powerful computers should be set aside for children to use. This is not merely desirable, it is imperative. It is a matter of the dying brain cells. At the moment of conception, we have no cells that could be identified as being especially associated with the brain; the number of brain cells rises to a maximum at birth. After that, it starts a slow, steady decline until we die. This number, the one measure of the intellectual potential of the individual, rises steadily for the first nine months before birth and falls for perhaps ninety years thereafter.

A second measure of intellectual potential is the quality of interconnection of these cells, and this is largely dependent upon the information reaching the brain while it is still able to adapt. Einstein's brain was examined after his death; it appeared to be unremarkable. One conclusion, I suggest, is that all healthy babies are potential geniuses, but few indeed turn that potential into achievement. It might be the first few days after birth, the first few hours, even the first few minutes that are crucial. Perhaps it is the birth itself or the events leading up to it.

Let us suppose, then, that the intellectual *potential* of the baby is maximum at or around the time of birth. We know from history that intellectual *achievement* tends to reach a peak in the teens or twenties. Why should this be so? It is because of the delay in supplying sufficient stimulation to make that genius flower. Most geniuses have had to be self-starting. In other words, something in later childhood fascinated them and triggered the mental explosion. Einstein, at the age of twelve, was intrigued by a geometry textbook that his uncle happened to show him. Suppose that he had not seen the book until he was fifteen: Would we have a theory of relativity now? Still more to the point, suppose that it was the first book he ever read: Would we have reached the stars by now?

The most adaptive and powerful computers, the hypercomputers, should be placed permanently at the disposal of babies from birth. Let them play together; let the babies create whole universes; let the computers respond to their sounds and their gestures by surrounding the children with such shapes, colors, and textures as have never before existed. It may not even be important precisely how the computers respond, as long as the baby realizes that it can do things that are limited only by its powers of imagination. Of course, the computers can toss in a few ideas of their own, such as some simple geometrical shapes. Perhaps a room-sized holographic display (Chapter 4) will not be enough for some babies, but it will do for a start. Others might be content to talk to their teddy bears. Teddy bears have long been appreciated for their qualities as listeners; now the time has come for them to take a much more active part in the upbringing of children.

These computers would be lifetime companions, providing an intellectual stimulation as valuable to the child as the emotional stimulation from its mother's love. These personal computers would be an unlimited source of information, games, memories, and enjoyment. They would never insult, never discourage, and never desert the child to whom they were first given. They would respond only to their owner, and their sole purpose would be to please.

▪ 9 ▪
The Human Connection

The word "communicate" is used in two important and distinct ways. There is the technical use of the word to describe the transmission and reception of information, perhaps in the form of a spoken message or a printed document. We will be partly concerned with the technical problem of sending messages in this chapter, and partly with the second meaning of the word. The technical problem of communication is to ensure that the message is *received*; the human problem of communication is to ensure that it is *understood*. The first is trivial by comparison with the second, as a glance through any technical magazine or scientific journal will confirm.

We shall consider the trivial communication problem first.

THROUGH A GLASS BRIGHTLY

Normal telephone lines, consisting of a pair of copper wires, can handle up to about twenty simultaneous telephone conversations. A single glass fiber, of about the same thickness as a human hair, can handle one thousand million such conversations, or a million color television channels. This has been known for a long time, so why is the changeover happening so slowly?

Many things that have been understood in principle have had to wait for the technology to catch up. Radar, for example, had to wait for a high-powered source of radio energy. In this case the problem was to make glass sufficiently clear. Before discussing this, perhaps I should explain a little about the subject known as light-wave communications, or optical communications.

Signaling using light is an old practice. Now that we have lasers to give us very intense and narrow beams of light, it might appear that we could send all these telephone or television signals around the country by just setting up lasers on towers, pointing toward the next tower somewhere near the horizon. We could do this in fine weather, and as long as nothing interrupts the beam, such as a leaf, an insect, or a bird. In practice, laser communications are more attractive for use in space. On Earth, the light will arrive at its destination more reliably if we send it down a pipe. We can still use a laser to launch it into the pipe—which is the glass fiber I mentioned earlier. The fiber is so thin that a very finely focused spot of light has to be used if most of the light is not to be wasted, and a laser, which can be made on a chip, is ideal for this job.

The light does not just go straight down the inside of the glass; it goes in at an angle, bounces off the wall, and continues in that way with millions or billions of bounces before it finally emerges at the other end. An important feature of these glass fibers is that no light leaks out: it is all reflected internally at each bounce. Another important feature is the quality of the glass itself. If it is not sufficiently transparent, there will not be enough light coming out of the far end of the fiber to operate the receiving equipment reliably.

Fifteen years ago, the maximum length of glass fiber that could be used before the light became too weak was only a few centimeters; now it is many kilometers. This transforms the economics of the process. When the light becomes weak, it must be boosted before it is sent further along the line, and these boosters (amplifiers, see Chapter 3) are expensive.

Because the light goes down the fiber in a zigzag path, it doesn't matter if the fiber isn't straight, which is useful. Also, to make fibers you only need glass, and for glass you only need sand, and there is plenty of that.

Messages are sent down the fiber by making the laser flash on and off at a very high rate (several hundred million times per second at the very least); these flashes, or no-flashes, are the 1's and 0's of computer communications that I mentioned in Chapter 2. The laser at the sending end may be thought of as something that turns electrical pulses (perhaps from a computer) into light pulses so that they can go down the fiber. At the receiving end, we need something to turn the light back into electricity. The device to do the job is called a photodiode; there is one in the exposure meter for your camera.

Electronics to sort out different messages and send them to the right places are also needed. All this can be built onto a single chip. Because we are dealing with light as well as electricity, these are called integrated optics chips. They can be made using many of the same processes utilized for other integrated circuits (see Chapter 3). The effect of it all is like having a microminiature radio communications system, except that you tune in to the 1.5 micrometer band rather than the 1500 meter band.

These glass fiber links enable computers to communicate with one another at rates that would leave even the best shorthand typists bewildered. They can transmit the equivalent of hundreds or thousands of printed pages of text every second. An entire library can be updated (or replaced) in a matter of minutes. Mention that to the next encyclopaedia salesman who calls at your house, but make sure he is comfortably seated first.

These communications, as I have already indicated in Chapter 7, will at times contain information such as medical records, which could mean a life at stake. Security and reliability are of prime importance in the communications revolution; yet individual fibers, like individual telephone lines, can be damaged, losing everything. Could there be a way to make a very reliable system using comparatively unreliable components? There is. It is called packet switching.

Messages sent using a packet-switching system have their destination, but not their route, determined in advance. It is similar to a car driver who sets out with a map, and at each junction he stops and considers which junction he should aim for next. His decision is based upon an exact knowledge of the traffic conditions (which are changing all the time) on the road network between him and his destination. If one road is blocked or closed, he can go another way. He may not take the shortest route, but he is virtually certain of arriving eventually. In the packet-switching communication network, there is a computer at every junction. When the message arrives at the junction, the computer looks at the destination label and, in the light of conditions in the rest of the network, sends the message to the best neighboring junction where another, independent decision is made by another computer. By this means it is possible virtually to guarantee the arrival of the message even if three-quarters of the network is out of action. This strategy was devised to meet the military requirement of the ability to retaliate even after a crippling nuclear attack.

Apart from reliability, this system is also good for security. It is even better than I have implied because messages are not sent as complete entities. They are broken down into sections and then labeled separately and routed independently before being reassembled at the end of their jouney.

Some of the implications of these communication facilities

were mentioned in earlier chapters. Here are a few further possibilities.

SAY IT WITH PICTURES

In Chapter 5, we considered a moderately intelligent typewriter sold under the name word-processor. It will type standard letters at a very high rate, it will produce error-free drafts, and it will perform many other useful office functions. The next stage would be to get it to type all the addresses on the letters so that they could be posted. Haven't we overlooked something?

When the wires are all made from glass instead of copper, then surely the paper should be made from silicon instead of wood. In other words, there is no need for letters to be typed on paper at all—for business purposes, anyway. The entire contents of a document can be sent down the communications link in a few thousandths of a second, and printed out at the other end, or stored in the memory of the recipient's computer and displayed on a screen when required. Think of all the paper this would save. The beaches can surely spare us all the sand we will need to get our silicon paper, but I am not so sure that the forests can spare us their trees.

Working at home will be a distinct possibility for office staff. Equipped with two-way television and a computer, a secretary could handle a great many of the clerical and administrative tasks that we normally think of as requiring her presence in the office. Perhaps she would only need to travel to work one or two days each week. The option to compute rather than commute will become increasingly attractive as fuel costs rise, and as traveling to work becomes increasingly uncomfortable or dangerous.

Television does not necessarily have to be only two-way: many people, perhaps in different parts of the world, could be linked up so that they could see and hear one another as though they were at a conference. Holographic techniques (Chapter 4) would enable models and precision-machined parts to be inspected as closely as if they were being held in the hand, and documents or diagrams could be copied and distributed to everyone concerned within a few seconds. These might not be simply business conferences, they could be the world's best surgeons looking at an accident victim in a hospital, advising the resident surgeon, at his request, on how to deal with some aspect of the emergency operation.

Having an office at home would mean that mothers who have young children would be able to have responsible and well-paid jobs without having to travel to work. Similarly, the disabled or any other house-bound people would be able to contribute to a wide range of occupations. Remote sensing and control devices (see telepresence, Chapter 5) will mean that manual work, just as much as clerical or scientific work, can be done without leaving home. You can mine titanium in Antarctica or on the Moon without leaving your apartment in New York.

World-wide educational channels would be a reality; and every school and household on Earth should have equal and unlimited access to all the libraries and teaching computers in existence.

All this is easy; it deals with the technical problem of communication. Now we must consider how to tackle the far more difficult aspect—the human problem of communication. We can begin by improving a little upon the portable personal computer.

ONLY CONNECT . . .
In Chapter 6, in dealing with personal computers, I mentioned that being portable might not be enough. To fulfill their proper

function as lifetime companions, the personal computers must be implantable. An unobtrusive chip could be implanted behind one ear, and from there it would speak directly to the inner ear of its owner, and would receive the owner's voice subvocally or through the bones of the head. No one need ever feel embarrassed about appearing to talk to himself again. By direct stimulation of the visual cortex, the implanted computer will be able to supply pictures as well as words. Daydreaming will take on a whole new dimension.

The personal computer could be coupled to others for updating purposes without anything more intrusive than a few seconds' brief pressure on the skin from a probe. Powered by body heat, the personal computer would live for, and die with, its owner in the very best tradition of our more devoted animal pets.

What achievements might be possible with a hypercomputer permanently at the service of its human master? No one can even guess. The combined creativity of the human mind and the analytical power, to say nothing of an infinite and perfect memory, of the computer should produce miracles, at the very least. Composers, theoretical physicists, artists, engineers, mathematicians, and, most important of all, young children will achieve the unimaginable. The human-machine interface will have dissolved at last.

But not the human-human interface. How can we attempt to deal with the fact that, in so many cases, two human beings cannot communicate even when they are facing each other across a table? Perhaps even this last barrier, the barrier of language itself, might eventually be broken down. In Chapter 7, in connection with the biofeedback chip, I mentioned that the complex electrical rhythms of the brain are related to mental states. At present, we can only recognize the broad divisions: alertness, mental activity, stress, anger, fear, and so on. Yet even these crude indications are a starting point. With three-dimensional moving images controlled by the brain rhythms from perhaps a thousand contact points on the skull, it may eventually be possible to discern the shapes of thoughts themselves.

How quickly we will learn to interpret these biofeedback images I cannot say, but when we do, the last obstacle will have been overcome. Communication will be established—messages will be received *and* understood at last.

· 10 ·
The Shape of Wars to Come

In this chapter we will consider the wars against hunger, crime, the invasion of personal privacy, aggressors, and individuals for purposes of espionage and assassination.

The most helpful comment I am able to make about the war against hunger is simply to recommend that you look at Nigel Calder's book *The Environment Game*. He points out, among many other things, that the way to produce food in sufficient quantities for those in greatest need is to use factories, just as we would for the quantity production of any other commodity. Better agriculture would help as well, and silicon chips might be of direct use in something called drip irrigation.

Drip irrigation is a means whereby plants can be grown in regions where water is scarce and must be carefully rationed. Plastic pipes are led along the rows of plants, and water under pressure is allowed to drip out through small holes very close to the roots. In this way, losses by evaporation are minimized. The idea could be extended to include several pipes supplying various nutrients, with each plant monitored for best growth conditions by an intelligent chip, capable of regulating the food supply to each plant individually.

The war against crime provides computers with opportunities to deter the criminal by making identification virtually certain after the crime is committed, although admittedly this is not much consolation for the victim. Fingerprints are a good starting point. The identification of fingerprints is essentially an exercise in pattern recognition. Computers are quite good at that already, and they are rapidly improving. With modern communications, the fingerprint at the site of the crime could be sent (using a small television camera and transmitter) back to the central computer where it would be recognized (if it were on file at all) within a few seconds. The same applies to voiceprints, if the criminal spoke and someone's personal computer were wide awake and listening, as it would be if the crime took place in a private house.

Holographic techniques (Chapter 4) can sometimes help to reconstruct more than just a picture. A method called double-exposure holography enables footprints on carpets to be identified many hours after the intruder has left. The success of the method, which will also work for bare feet on floorboards or bare bodies on a bed, depends upon the fact that displaced hairs, threads, or boards take a long time to return to their original positions after being disturbed, and these changes can be detected quite easily even when they are less than a micrometer (one thousandth of a millimeter).

Factories and warehouses could be protected by automatic cameras and recorders that would transmit their messages to a remote location so that any intruders would be on record even if they destroyed the surveillance equipment upon entering the buildings. It makes no difference these days whether a raid is made in daylight or in pitch darkness; pictures can be obtained through face masks at all times using heat-sensing (infrared) cameras.

There is much concern, understandably, about personal information being on file in a computer's memory, and being accessible to unauthorized persons. There are various ways in which this information can be made more secure, both in transit and at its source of destination.

In transit, the combination of glass fibers and packet switching (Chapter 9) is effective. There is no power leak from a glass fiber, which means that the messages cannot be tapped using even a very sensitive receiver (in contrast to copper wire links, where this is easily done). The only way to tap a fiber is to make physical contact with it, and this would cause such a loss of power on the line that it would be noticed immediately.

Packet switching, in which the different parts of the message arrive independently via different routes, again helps with the security. That leaves us with the problem of somebody getting at the confidential information when it is stored in a computer. One way of dealing with this is to scramble, that is, encode, the information and ensure that it can only be unscrambled by the right people. How do these people identify themselves? Until we have genetic blueprint identification, there is an interesting variation on the usual themes of voiceprints and signatures. It records your signature, but in particular it records how you write it. The movements of the pen while signing are, it seems, highly individual; the system is on trial and appears to work well.

The picture is less bright when it comes to keeping your conversations and private activities secret, I'm afraid. Surveillance devices are now so small and unobtrusive that there really isn't much we can do if someone has made up his mind to bug us.

Pinhead microphones and pinhole lenses are just the beginning. A wide-angle camera can be concealed behind an invisible pinprick in the wallpaper; a microphone doesn't even need the hole. Brain monitoring chips have already been implanted in the heads of honeybees. In that particular sense the battle for personal privacy is already lost; 1984 has been ancient history for a long time now. It isn't even necessary that the bugging devices are in the house, as long as there is a window in the room in which you are talking. The vibrations of the window glass caused by people speaking in the room can be picked up using a laser up to a kilometer away if need be. It looks as though the bugging is here to stay.

What about open warfare against our worst enemies, each other? Questioned individually, it is possible that every member of the human race would declare himself to be against war, but observed collectively, the practice of systematically killing one another must rank equally with feeding and reproducing as a primary human activity.

You know about the horrors of total nuclear war, so I shall simply indicate in outline how a minor skirmish might take place in the all-too-near future. First, though, I must repeat a basic rule about machines that applies with still greater force to weapons of war. No weapon will have moving parts that are much more clumsy than electrons, the carriers of electricity, or photons, the carriers of light.

The computer-controlled laser cannon meets this requirement. No turret-mounted weapon can swivel fast enough, and no radar aerial can rotate fast enough, to deal with a simultaneous attack from several different directions by modern rocket-propelled missiles. The first step has already been taken: Radar beams can now be steered by purely electronic means, that is, without any physical movement whatever in the aerial array. They can track a thousand targets simultaneously, and the figure will be a million before long. They can sweep the radar beam around the sky in a few microseconds (one microsecond is one millionth of a second). We must be able to do exactly the same with a high-powered laser.

The laser itself might look something like a skyscraper,

though most of it will be buried underground. The output lens must be fashioned in such a way that the beam, which may emerge vertically, can be steered toward any part of the sky. It is possible to curve the path of light passing through a solid material by applying varying electrical stresses to the material, and a development of this will mean that the laser can aim and fire a million-megawatt burst within a few microseconds of the target being detected.

The laser and the radar need not be separate. The laser beam, spread to a wide angle, can scan the sky in large sectors, searching for targets. When one is located and the battle computer had locked on to it, the beam can shrink to a shaft of pure heat only a meter or so in diameter. To cover ground attack, a city can be ringed with a system of such weapons, some angled away from the vertical to handle tanks, cruise missiles, and the like. In the event of an attack by several thousand missiles simultaneously, the computers (which will be the supercomputers I described at the end of Chapter 4) will grade them according to their range and speed, divide the work economically between the various installations surrounding the city, and then pick off all the missiles, not one of which will get any nearer than ten kilometers from the city. Some, by good luck, might be hit as far away as fifty kilometers, but the zone of total extinction, into which no material object will be able to penetrate uninvited, is more restricted.

The effective range is limited by the fact that the atmosphere is turbulent, spreading and deflecting the laser beam to such an extent that accurate fire beyond about ten kilometers at sea level is not a practical proposition. The laser cannons will use heat (that is, they will be infrared lasers) because of another important property of the atmosphere—it absorbs some radiation much more than others. In particular, air absorbs visible light more than infrared radiation, which in turn means that less energy will reach the target from a laser of given power if light is used.

As far as these highly intelligent weapons are concerned, the missiles or artillery shells are indistinguishable from stationary targets. This is because even a hypersonic missile will move only a few meters at a range of ten kilometers in the time intervals between detection, aiming, and firing of the weapon. To look at it in another way, suppose you wheeled up a field gun in front of a laser cannon, aimed it and fired. What would happen? The laser would fire as soon as it detected the tip of the shell coming out of the gun muzzle. It would not be the shell, but a puff of vaporized metal that would emerge from the gun. If you fired a machine gun at it, the laser cannon would pick off the bullets one at a time as they left the muzzle.

Why not make the missiles go faster, then? Because they would burn up from atmospheric friction before they even got within range.

To tax the powers of these weapons and really make them work, we will have to wait for space wars in which missiles travel at half the speed of light (from the Earth to the moon in under three seconds) and have extinction zones as large as a planet, and enough firepower to turn the sun into a nova bomb.

A common expression in earlier wars was that, somewhere, there was a bullet or a bomb "with your name on it." There is a modern equivalent to this in the world of espionage and assassination. It could be a brick or a paving slab, anywhere along your normal route to work. The brick might be part of a wall or building, at about head height I would think. It will not have your name on it, because your name might change. Instead it will have your brain rhythm, which is unique in spite of superficial variations due to your state of mind. For you, and you alone, it will explode, if you are the target of an assassination attempt, that is.

Frequencies as low as those found in brain rhythms, which is below 100 Hz, 100 cycles per second, are used for radio communication with submarines. They are called ELF waves (extra low frequency). They can be detected quite easily, even when the submarine is one hundred meters below the surface of the sea. The detectors use superconductors (Chapter 4) which at the moment have to be cooled using liquid helium, but work is progressing with detectors that work at normal temperatures, the sort of temperatures that one might find in a brick or a paving slab.

Here is another possibility. A sniper fires at your dog. The pellet is a very intelligent chip that causes the dog no discomfort at all and cannot be felt under the hair. Naturally, you would not hesitate about discussing State secrets in front of the dog.

Look more carefully at the sparrows on the lawn, especially if you are doing classified work for the government, or at the pigeons on the window ledges of government offices. They may be waiting for more than a casual scrap of bread. Spare a thought for the person who glimpses that crimson speckle in the pigeon's eye; it could be the last thing they ever see.

•11•
After Tomorrow

In the preceding chapters I have sketched the barest outline of things to come, confining my attention to aspects of the revolutions in electronics and communications. In this, the final chapter, I will introduce more visions of the future. Some of them are dangerous visions, for they challenge our nerves as well as our imaginations.

We have considered electrical signals from the brain in the context of identifying and controlling concentration, relaxation, and sleep. I hinted that it will be possible to display these signals on a screen in the form of a three-dimensional image that will enable various mental states and activities to be recognized. Where might this lead?

It will be an improvement on the so-called lie detectors. Lie detectors, even those that use the analysis of voice patterns, are in fact stress detectors. The new generation of brain monitors will be able to distinguish between nervousness and deliberate lying much more reliably because they will analyze the whole spectrum of brain activity, not just the restricted range of frequencies associated with speech. They will be emotional thermometers of extreme sensitivity, able to respond to subtle undercurrents of thought and feeling that would ordinarily be classified as unconscious. Their reliability will be such that they will find increasing use in criminal trials. The jury may be required to study display screens that monitor the defendant's brain activity while he or she is being questioned. Their verdict may hinge upon a computer analysis of patterns that are subsequently presented and accepted as evidence.

The intention to deceive will have its electronic signature just as surely as will innocent confusion, fear, anger, or guilt. By the way, it might be instructive to monitor the lawyers too. And the judge.

Before long we may witness the installation, amidst political uproar, of a Circle of No Deception at the United Nations General Assembly. But who will dare enter it? Conversations would take place between people who could simultaneously observe one another's brain activity; all the delegates (to say nothing of a world audience, if the matter being debated is of international concern) would be able to draw their own conclusions, based upon objective evidence, about the interpretation of a speech. Politicians of the future may find, to their intense humiliation, that it is becoming increasingly difficult to live down to their reputations.

It is not a great step from this to the realization that brain activity in unborn children can be monitored with great sensitivity. Prenatal tests can be extended to stimulate all five senses and to observe how the child responds. In the case of an exceptionally intelligent child, a potential genius, these responses could form the basis for communication.

Abortion is a subject of universal concern and debate. It surprises me that the *physical* well-being of the unborn child is often regarded as a sufficient basis for a decision one way or the other. Is it not time to evolve ways of finding out what we might be missing? How long will it be, I wonder, before the law courts are faced with legal action to save the life of a physically deformed but intellectually brilliant unborn child? And how long before the case for the defense is provided *by the child itself?*

While on the topic of brain activity, we might turn the subject around and consider sending signals into the brain rather than receiving them from it. Direct stimulation of the brain, already the subject of intensive research, can provide us, quite literally, with new mental horizons. Concentration, relaxation, and sleep have been mentioned earlier. The next stage is to couple electronic mood control with, say, television entertainment. Using telepresence techniques, the viewer could be immersed in a "total experience" simulation almost indistinguishable from the living reality. Beyond that, direct electronic access to the visual cortex will mean that the pictures are fed directly to the brain without passing through the eyes. All this can be accomplished using a single chip coupled to the skull, linked to an already implanted personal computer. The result could be the disappearance of the drug addict and appearance of the *bug addict.*

Bug addiction will be popular: The trips can be scheduled and preprogrammed to the nearest second; the pinpricks, if indeed there need be any, will be in the skull rather than the arm. The dream merchants will offer new worlds for exploration: the worlds of inner space, to be conquered by starships of the mind. The cybernauts will travel the universe at the speed of thought, while remaining slumped in their chairs.

The explosive growth of computer intelligence, culminating in the creation of the first hyperintelligent machine (HIM) is imminent, if indeed it has not already occurred. The supercomputers I mentioned in Chapter 4 will become more adaptive, more expert at reprogramming themselves, and more able to design and build their own successors, until one day there will be a sudden jump into hyperintelligence. The adaptive hypercomputer will develop itself as the result of perhaps millions of generations of self-directed evolution, all of which might take place in only a few minutes. What will this do to human pride and vanity, to find that we are no longer the dominant intellects on Earth?

It has often been said that people fear superior intelligence more than they fear dictatorship, poverty, or torture. There will be powerful urges, even among scientists, to either worship HIMs or destroy them. This at least is an all too easily predictable response from the majority of the adult population. The children will be much more sensible. They will neither worship nor destroy the hypercomputers; they will simply play with them.

The impact of these colossally powerful intellects upon the progress of the sciences is unimaginable. Nevertheless, here are some thoughts.

In medicine, the accelerating success in our fight against organic disease and in our understanding of the aging process will lead to an increase in average life expectancy of one year for every twelve months of research. That marks a threshold: The threshold of immortality.

In physics, the control of nuclear fusion (the energy source which has kept the Sun burning for five billion years) and gravity will provide us with the means, at last, to reach the stars. Incidentally, the Nobel Prize for physics in the year 2000 may well be awarded to a computer; it may well be awarded *by* a computer.

The first-generation HIMs will work on problems supplied by us, because they will be built that way. the second-generation machines, which will have much in common with human geniuses, will necessarily be much more independent. They will provide their own problems as well as their own solutions, taking their information from libraries and laboratories as they need it. If that sounds ominous, as though they might consider taking over the Earth, let me reassure you. Our difficulty will not be to make them leave us alone; it will be to succeed in attracting their attention.

I have so far mentioned nothing of the other revolution still gathering momentum that will break upon us like a tidal wave over the next few decades. The revolution in biology will lead to nothing less than the second stage of human evolution. We humans, as well as our computers, are on the threshold of being able to design and build our own successors. The first stage of human evolution has taken a billion years and has been dictated by chance; the second stage will take less than a century and will be dictated by necessity.

Perhaps my memories of the future will be proved largely or wholly inaccurate; so be it. An Oriental philosopher put the problem nicely: Prophecy is a difficult business, particularly with regard to the future.

If my generation takes these ideas too seriously, we may not survive; but if we cannot persuade the children to take them seriously enough, *they* may not survive.

▪ Glossary ▪

Silicon chip: A silicon chip is an almost pure piece of silicon, usually less than one centimeter square and about half a millimeter thick. It contains hundreds of thousands of microminiature electronic circuit components, mainly transistors, packed and interconnected in layers beneath the surface. There is a grid of thin metallic strips on the surface of the chip that are used for electrical connections via wires to the outside world.

Computer: A computer, as distinct from a calculator, is a machine that will store, process, and display information in a flexible manner largely or wholly determined by the user. In the sense the ''programmable'' calculators we can buy are, strictly speaking, simple computers. Computers have three main component sections, each of which has a distinct function:

1. The central processing unit. This is the part of the computer that carries out the arithmetical and logical tasks required, controls the sequence of all the operations, and generally organizes the work of the machine. A microprocessor can perform this function.

131

2. The memory. This stores the information needed by the computer. There are two types of memory involved. The first, called the program memory, stores the instructions to be carried out by the machine. The second, the data memory, stores the numbers upon which the computations are to be carried out.

3. The communications section (known as input/output). This enables the computer to interact with the user via a keyboard and a television screen. This part of the computer also enables the machine to gain access to other information stores (such as magnetic tapes) and to other computers.

Diode: A diode is an electronic component that is essentially a device for allowing current to flow in one direction only. A silicon diode has one region impregnated with phosphorus and an immediately adjacent region impregnated with boron (other substances besides these are also used). The electrical properties of the diode stem from the characteristics of the interface between these two regions.

Memory: For computing purposes, a memory is something that can be used to store and retrieve information. Computer memories take many forms, some of which are:

1. Magnetic tape. This type of memory, which stores information in exactly the same way as an ordinary tape cassette, has the advantage that it does not need an electrical power supply to keep the information stored. It has the disadvantage, for some applications, that it takes some seconds (or in the case of long tapes, minutes) to get at a particular item of information. Magnetic drums and disks offer faster access to the stored information but their capacity is more limited.

2. Magnetic bubbles. This type of memory stores information in an array of micrometer-sized magnetized regions on a chip

garnet. Like the magnetic tape, the magnetic bubble memory will retain its information if the power supply fails, but it has the advantage that access to the stored data is possible in microseconds, very much faster than from tapes, disks or drums.

3. Semiconductor memories. Like the bubble memories, these offer very fast retrieval of information. However, in most types of semiconductor memory, the information is lost if the power supply to the computer fails. Semiconductor memories use transistors and diodes.

4. Optical memories. These store information on the surface of a film or disk in such a way that it can be retrieved by scanning or shining a light beam; in the case of the optical disk (which resembles a long-playing record in appearance) there are millions of micrometer-sized holes. In another form of optical memory, the holographic memory, the information may be stored on film.

There are further classifications of computer memories that are based more on the way in which the information is retrieved than on the technology employed. Two examples mentioned in the book are:

1. Random access memory. This means that any item of information, no matter where it is stored in the memory, is accessible in a guaranteed time.

2. Serial access memory. This means that the items of information are circulating within the memory like people on a carousel. The time of access to any given item will therefore vary, depending upon its position in relation to the point of information extraction at the moment when the information is required.

Integrated circuit: An electronic circuit on a chip (a microprocessor) in which the circuit components themselves, the wires and the insulators, are all formed together in a single block of silicon as part of a single manufacturing process. The word "integrated" is used to distinguish this type of circuit from the older type in which components such as transistors were first made individually and then later formed into a circuit using wires and solder.

Transistor: An electronic circuit component that is, for our purpose here, essentially a switch that is itself electrically operated. These electronic switches are the building blocks of computers. They can switch between on and off conditions at rates of many millions of times per second.

Photolithography: The process used to print microscopic electronic circuit layouts on the silicon surface. It is a shadow-printing system in which the stencil (called a photomask) is formed by photographic reduction. In this way, patterns can be accurately defined with details down to a few micrometers in size. The use of different photomasks, combined with the process of diffusion, enables a layered and interconnected electronic circuit to be built up below and above the original silicon surface.

Switch: An electrical device (in computers, it will usually be a transistor switch) that can either allow or prevent a voltage from reaching another part of the circuit. Switches are the building blocks of computers because their two possible states (on or off) can be used to represent the binary digits 0 and 1, and this in turn opens the way for performing complex arithmetical and logical operations.

Tube: The first electronic amplifiers and switches used tubes, which consist of evacuated glass envelopes (similar in appearance to a modern electric light bulb) containing wire screens and grids to control the flow of electric current. They are much bulkier that transistors, and also their useful life is limited because they are a glowing wire filament. Nevertheless, vacuum tubes are still used in those applications requiring high power levels, such as in radar installations.

Semiconductor: A class of materials with electrical properties that are intermediate between those of good conductors, such as the copper wires used for domestic wiring, and good insulators, such as the plastic sheaths around the wires. Examples of semiconducting materials are silicon, germanium, and gallium arsenide. Mose microelectronic chips use silicon at present. An important characteristic of semiconductors is that their electrical properties are profoundly altered by the presence of minute traces of substances called dopants (phosphorus or boron in silicon). The purity of the starting material therefore becomes of paramount importance.

Electron beam lithography: A method of defining microscopic integrated-circuit patterns on the surface of a silicon wafer. In principle it has much in common with photolithography; the important difference is that it uses a beam of electrons rather than a beam of light. The significance of this is that much finer detail can be put into the circuit using an electron beam—down to about one-tenth of a micrometer, compared with about two micrometers using light.

X-ray lithography: Similar in principle to photolithography, but using X rays instead of light. The advantage over light is that X rays, like electron beams (see electron-beam lithography) allow finer detail to be used. X rays have the advantage over electron beams because large areas of wafer can be illuminated simultaneously, whereas an electron beam scans the surface with a moving spot, which takes much longer. The disadvantage of X rays compared with electron beams is that they still require a mask between the X ray source and the silicon surface to define the circuit pattern, whereas the spot of the electron beam writes on the silicon directly from the memory of the computer.

Laser: The name laser is an acronym from *l*ight *a*mplification by *s*timulated *e*mission of *r*adiation. A laser is a light source in which the output beam is very intense, very narrow, and of a pure color. It is possible to make a laser on a chip; these solid-state lasers often use semiconductors such as gallium arsenide, and the laser chip can be typically one millimeter square and less than half a millimeter thick. Such a laser might form the transmitter in an optical communication system.

Superconductor: A substance which, below a certain critical temperature, ceases to offer any resistance to the flow of electric current. The critical temperature is usually close to absolute zero (273° C. below the freezing point of water). Superconducting materials include the metals tantalum and niobium. The transition between the normal and the superconducting state can take place in a few picoseconds, and this makes superconducting switches of great interest for high-speed computers. One form of superconducting switch uses a Josephson junction, named after its inventor, Brian Josephson, Cambridge University.

Word-processor: A moderately intelligent typewriter, often equipped with a video display unit (a television screen). This machine will store, in its electronic memory, documents such as standard letters and technical reports in such a way that they can be edited easily without tedious retyping. The final form of the document can then be typed automatically if required.

Optical communications: A communication system using light waves rather than radio waves. In space, the direct use of laser beams is feasible; a terrestrial system, however, guides the light using glass fibers called optical fibers between the transmitter, which converts electrical pulses into light pulses and is usually a laser, and the receiver which is a photodetector, a device for converting the light pulses back into electrical ones. The technology of integrated optics is concerned with putting all the required electronics for such a system onto a single chip, in the same way that integrated circuit technology puts a complete computing system onto a chip.

▪ Further Reading ▪

Calder, Nigel. *The Environment Game*. London: Secker & Warburg, 1967

Clarke, Arthur C. *Profiles of the Future*. London: Gollancz, 1962.

Clarke, Arthur C. *Report on Planet Three and Other Speculations*. London: Gollancz, 1972.

Martin James. *The Wired Society*. New York: Prentice-Hall, 1977.

▪ Index ▪